T0135606

Simulation of steel quenching with interaction of classical plasticity and TRIP – numerical methods and model comparison

von Bettina Suhr

Dissertation

zur Erlangung des Grades eines Doktors der Ingenieurwissenschaften

–Dr.-Ing.–

Vorgelegt im Fachbereich 3 (Mathematik & Informatik)

der Universität Bremen

im März 2010

Bibliografische Information der Deutschen Nationalbibliothek

Die Deutsche Nationalbibliothek verzeichnet diese Publikation in der
Deutschen Nationalbibliografie; detaillierte bibliografische Daten sind
im Internet über http://dnb.d-nb.de abrufbar.

ISBN 978-3-8325-2519-4

Logos Verlag Berlin GmbH
Comeniushof, Gubener Str. 47,
10243 Berlin
Tel.: +49 (0)30 42 85 10 90
Fax: +49 (0)30 42 85 10 92
INTERNET: http://www.logos-verlag.de

Abstract

This thesis deals with the mathematical modelling and numerical simulation of processes involved in the quenching of steel. The considered physical processes are heat conduction, phase transformation, thermoelasticity, classical plasticity and transformation induced plasticity (TRIP). Although both modelling and simulation are addressed the scientific contribution of this work lies in the field of numerics.

This work is motivated and supported by the Collaborative Research Centre SFB 570 "Distortion Engineering - Distortion Control in the Production Process" of the German Research Foundation (DFG). In several situations, experimental results and simulation results, which were obtained using wide-spread standard models, are in no good accordance. This is the starting point of this thesis, it tries to reduce the deviation between experimental and simulation results by using different or more complex models. In a first step relatively simple experiments are used to determine all necessary model parameters (verification step). Then the models are applied to more complex settings to see, whether the simulation results are closer to the experimental ones (validation step).

In some cases not all experimental data necessary for the verification and validation of a model are available. For the verification we try to derive the necessary model parameters from comparable materials or, if this is not possible, we vary the parameters. Where validation experiments are missing, we compare different models (possibly with parameter variations) to investigate their influence on the simulation result.

The processes for which different models are compared are phase transformation, transformation induced plasticity and the hardening behaviour of classical plasticity. In the numerical treatment of steel quenching, the main challenge arises from the mechanical behaviour. Besides the systematical comparison of different models, the major contribution of this work is the integration of classical plasticity into an already existing numerical scheme for thermoelasticity with TRIP. For the classical plasticity two models of combined isotropic and kinematic hardening are compared. In both hardening models an interaction with transformation induced plasticity is taken into account. All simulations are conducted with the open source FE-toolbox ALBERTA.

The systematical model comparison is a suitable method of determining the influence of a physical process on the simulation result. In this way it can be investigated for which physical processes additional experiments could help to improve the simulation quantitatively.

Contents

CHAPTER

1

Introduction

This thesis is positioned in the overlap of two scientific disciplines: applied mathematics and materials science. It deals with the mathematical modelling and numerical simulation of processes involved in the quenching of steel and was written within the Collaborative Research Centre SFB 570 "Distortion Engineering - Distortion Control in the Production Process" of the German Research Foundation (DFG). Although both modelling and simulation are addressed, the scientific contribution of this work lies in the field of numerics.

Steel is an extensively used material all over the world. Distortion arising in the production of workpieces is very expensive to remove in post-processing. Therefore it is of major interest to understand in detail the causes of distortion in each production step. The modelling and simulation of essential effects can help to gain a deeper understanding of the material behaviour.

In this thesis we will focus on the quenching of steel. There are several effects involved which interact with each other: During the cooling of steel a solid-solid phase transition takes place, which releases the so-called latent heat. The mechanical behaviour can be split into three categories: elasticity, classical plasticity and transformation induced plasticity (TRIP). The effect of creep is not considered in this work. The density of the

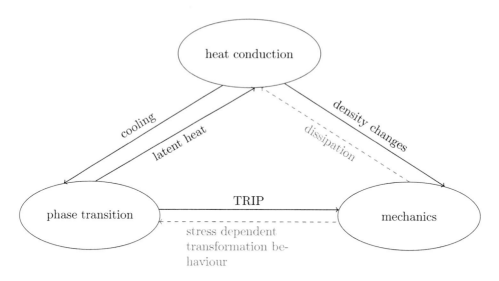

Figure 1.1: Scheme of the processes during the quenching of steel and their interactions, diagram taken from [INKM81]

material depends on both temperature and phases and particularly the phase change influences the mechanical behaviour directly via the transformation induced plasticity.

Not taken into account are the following interactions: the dissipation process where mechanical energy is transformed into thermal energy and the stress dependent transformation behaviour where the mechanical behaviour influences the phase transition. The interaction between all processes is sketched in Figure 1.1.

For some of the above described effects which occur during steel quenching there exist various models. In several situations, experimental results and simulation results, which were obtained using wide-spread standard models, are in no good accordance. The aim of our research is to reduce the deviation between experimental and simulation results by using different or more complex models. We will deal with the wall bearing steel 100Cr6 (SAE 52100) which is subject of intense research within the SFB 570, so that there is a lot of experimental data and material parameters available. All simulations were carried out with the open source Finite Element toolbox ALBERTA, which was developed at the universities of Freiburg, Augsburg, Bremen and Duisburg, see [SS05]. As this software is open source it is possible to implement different models, which is often restricted in commercial FE code. The material parameters used for the simulations are taken from [ADF+08a, ADF+08b] and are for convenience also repeated in Appendix B.

Besides a systematical comparison of different models for phase transformation and transformation induced plasticity which is a continuation of the work in [SFHW09], the major contribution of this thesis lies in the numerical treatment of the mechanics of the problem. For the classical plasticity there are two hardening models introduced for combined isotropic and kinematic hardening which both include a coupling with transformation induced plasticity. A numerical treatment of such models can not be found in the literature. For both models a semi-implicit numerical scheme is derived. In order to identify the model parameters for the more complex hardening model an optimisation procedure is developed (together with M. Wolff, University of Bremen). Although in this setting there is no experimental data available for a direct comparison to simulation results, it can be seen that the different models including a different coupling to transformation induced plasticity cause considerable deviations in the simulation results. Therefore a more thorough experimental investigation of this effect seems to be sensible. Such experiments are planed in a spin-off project of the SFB 570: "Mehr-Mechanismen-Modelle: Theorie und ihre Anwendung auf einige Phänomene im Materialverhalten von Stahl" (BO 1144/4-1), which is a project of the German Research Foundation (DFG). The interaction between classical plasticity and transformation induced plasticity is a concrete application of a two-mechanism approach (cf. [WBH08]).

1.1 Remarks on steel

Steel is an extensively used material and there is a wide range of standard literature dealing with steel, its properties and its use. In this section we will restrict ourselves to an explanation of the structure of steel, the processes during phase transformations and conclude with some short remarks on grains, lattice imperfections and alloying.

As steel consists mainly of iron and carbon, we will start this section with some explanations on the iron-carbon system. For brevity we give a short overview on the processes of phase transformation only and refer the reader to [Hor92, Mac92] for details.

In general, carbon can occur as graphite or as iron carbide (also called cementite) within the iron-carbon system. The case where iron and cementite (Fe Fe_3C) form a metastable system includes the materials steel and cast iron. Steel has a carbon content between 0.02% and 2.06%, while cast iron has a carbon content from 2.06% to 6.67% (in pure cementite, Fe_3C, there are trice as many iron atoms than carbon atoms, using the masses of the atoms this yields 6.67%, which is the maximal carbon content in the metastable iron-carbon system.)

In Figure 1.2 an iron-carbon diagram is shown. Choosing a carbon content one can see

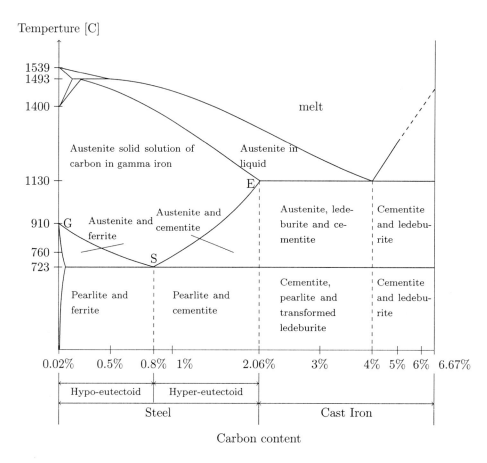

Figure 1.2: Iron-Carbon diagram.

which phase transitions occur during slowly cooling. It is important to mention, that in this diagram the system is assumed to be in equilibrium. Therefore the non-equilibrium phase martensite, which forms only in quick quenching processes, is not present in this diagram.

For hypo-eutectoid steel, which has a carbon content lower than 0.8%, the austenitic phase starts to transform to ferrite when the temperature drops below the line GS. Below the temperature 723°C the leftover austenite transforms into pearlite[1]. Steel with a carbon content of 0.8% is called eutectoid and below the temperature 723°C the austenite transforms completely into pearlite. For hyper-eutectoid steel with a carbon content

---

[1]Although pearlite is a two-phased lamellar structure, which consists of ferrite and cementite, we will refer to it as phase.

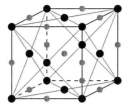

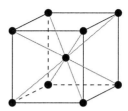

(a) Austenitic lattice structure: Iron atoms (black) form a face-centered cubic lattice and possible carbon atoms (gray) are positioned on the middle of the edges and in the center of the cube

(b) Ferritic lattice structure: Iron atoms (black) form a body-centered cubic lattice

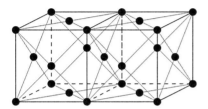

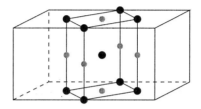

(c) Martensitic deformation: left austenitic lattice, right body centered tetragonal lattice of martensite with possible carbon atoms (gray)

Figure 1.3: Lattice structure of austenite, ferrite and martensite.

between 0.8% and 2.06%, below the line SE there forms cementite and below the temperature 723°C the leftover austenite transforms into pearlite.

For a deeper understanding of these phase transitions it is essential to know that steel is a polycrystalline material. It consists of grains, whose crystallographic orientations vary. The austenitic phase is a mixed crystal (solid solution) where the iron atoms have a face centered cubic lattice structure in which carbon atoms are possible in the middle of the edges and in the center of the cube, compare Figure 1.3(a). If the austenite lattice is completely filled with carbon atoms, then the material has an overall carbon content of 2.06%.

For hypo-eutectoid steel austenite is stable if the temperature is above the line GS. Below it, the lattice structure of the iron atoms start to change to a body centered cubic lattice: ferrite, see Figure 1.3(b). As the maximal carbon content in ferrite is 0.02%, the leftover carbon atoms diffuse into the austenite lattice, increasing the carbon content in the austenitic phase. The process of cooling and carbon diffusion proceeds until at 723°C a carbon content of 0.8% in the austenitic phase is reached and the leftover austenite transforms into pearlite. For hyper-eutectoid steel below the line SE there starts the

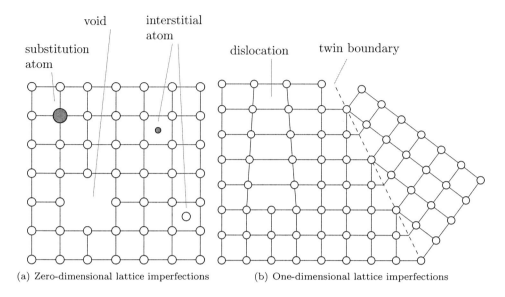

(a) Zero-dimensional lattice imperfections   (b) One-dimensional lattice imperfections

Figure 1.4: Lattice imperfections at one and two dimensions.

formation of cementite. Cementite has a relatively complicated lattice structure. In the orthorhombic elementary cell there are twelve iron and three carbon atoms. Due to this transformation the austenitic lattice looses carbon atoms. With the cooling proceeding and the carbon content in the austenitic phase decreasing, a carbon content of 0.8% in the austenitic phase is reached at 723°C and the leftover austenite transforms into pearlite.

The above described transformations are equilibrium transformations, where the carbon atoms have time to diffuse into the new lattice structures. Non-equilibrium phase transitions, which are caused by quick quenching processes, can not be described with the iron-carbon diagram in Figure 1.2. One important case of non-equilibrium transformations is the forming of martensite in quick quenching processes. Here the austenitic face centered lattice changes instantly into the body centered tetragonal lattice of martensite, while the carbon atoms stay at their places, compare Figure 1.3(c). Please note, that the lattice parameters of austenite and martensite are not equal.

It is important to mention, that within every crystal there are lattice imperfections. We will see later in Section 2.3.3, that the one-dimensional lattice imperfections called dislocations play an important role in the modelling of classical plasticity. In Figure 1.4 there are shown some zero- and one-dimensional lattice imperfections. Details on this topic can be found in the standard literature or in [Mac92].

Steel is alloyed with other materials to adapt its properties to the variety of its intended use. Examples of properties which are enhanced are tensile strength, corrosion resistance, ductility, wear and weldability. Also the grain structure can be affected, so that it becomes coarser or finer. Frequently used materials for the alloying of steel and their effect can be found e.g. in [Weg04]. In this work we are dealing with the wall bearing steel 100Cr6, which consists of the following constituents.

C	$0.93\% - 1.05\%$
Si	$0.15\% - 0.35\%$
Mn	$0.25\% - 0.45\%$
P	$< 0.025\%$
S	$< 0.015\%$
Cr	$1.35\% - 1.60\%$
Mo	$< 0.10\%$
Cu	$< 0.30\%$
Al	$< 0.05\%$

Table 1.1: Chemical composition of 100Cr6 steel (SAE 52100) as defined in EN ISO 683-17 in mass-%

1.2 Chapters overview

This work is organised as follows: Chapter 2 starts with the modelling of the different processes which are involved in our problem setting, the quenching of steel. For simplicity we introduce in the first three sections the modelling of heat conduction, phase transition and deformation behaviour (elastic, plastic and transformation induced plastic) without their interactions. In the fourth section we gather the equations for the single processes and introduce the coupling terms which are considered, thus obtaining the fully coupled problem. The following section deals with the weak formulation of our problem and in the last section we present some remarks on the lacks of knowledge on existence and uniqueness of the solution of the problem.

In Chapter 3 we will explain how the problem, formulated in the second chapter, can be solved numerically. First we will point out briefly that there are different approaches for dealing numerically with the deformation equation. In the second section we will present an overall numerical solution scheme for our problem. The following three sections will describe in detail the solution of the different types of equations: the application of the Finite Element method to the partial differential equations for heat conduction and the deformation equation, the numerical solution of ordinary differential equations for phase transitions and transformation induced plasticity and a detailed solution procedure for the problem of classical plasticity. The numerical treatment of this particular model for classical plasticity together with a coupling to transformation induced plasticity is a new contribution and can not be found in the literature in this form. In the sixth section we describe how periodic and symmetric boundary conditions can be included into the implementation. The last section deals with the a posteriori error-indicators which are used for the adaptive Finite Element computations.

After the mathematical modelling and the development of a numerical solution scheme, we will present in Chapter 4 the corresponding simulations. Results of simulations using wide-spread standard models are often in no good accordance with experimental results. The aim is to reduce the deviation between experimental and simulation results by using different or more complex models and conducting a systematical model comparison.

Starting with some preceding remarks on the difficulties in interpreting experimental results in the first section, we will deal with the martensitic transformation and transformation induced plasticity in the second section. In the third section we will discuss, whether the inertia term in the deformation equation, which represents oscillation behaviour of a specimen, can be neglected for our applications. The last section deals with a model comparison for classical plasticity. We consider two different hardening models for combined isotropic and kinematic hardening which both include a coupling with the transformation induced plasticity. The model parameters of one hardening model are available within the SFB 570. For the other hardening model there is derived an optimisation for the determination of the model parameters from experimental data (this was done together with M. Wolff, university of Bremen). There are comparative simulations done, which show the different behaviour of both models, the effect and the extent of the coupling between classical plasticity and transformation induced plasticity. Finally a simulation of an asymmetrically quenched ring shows qualitatively accordance with experimental measurements.

This thesis will be concluded with an outlook in Chapter 5.

1.3   Acknowledgements

This work could not have been composed without the the support
of several people. At first I would like to thank Prof. Dr. Alfred
Schmidt for the supervision of this work and his guidance, as well
as Dr. habil. Michael Wolff for his advice and many fruitful discussions.

Moreover I like to acknowledge the German Research Foundation
(DFG) for the funding of my work and my colleagues at the Collaborative Research Centre SFB 570 "Distortion Engineering - Distortion Control in the Production Process" for providing plenty of
material parameters and experimental data and for many helpful
discussions.

I would also like to thank the University of Bremen and the Zentrum
für Technomathematik for supporting my work.

Finally my deepest gratitude belongs to my friends and colleagues
who encouraged and supported me while writing this PhD-thesis.

Bettina Suhr

CHAPTER

$$2$$

Problem setting

In this chapter we will introduce in detail the setting which will be considered throughout this thesis. As we are aiming to simulate the quenching of a steel specimen, we will start with the modelling of all involved processes: heat conduction, phase transition and deformation behaviour: elastic, plastic and transformation induced plastic.

The first three sections describe the modeling of these processes (in the above order). For simplicity their coupling is neglected initially. In the fourth section we gather the equations for the single processes and introduce the coupling terms which we consider, thus obtaining the fully coupled problem. In the fifth section we present the weak formulation of our problem and conclude in the last section with some remarks on the lacks of knowledge on existence and uniqueness of the solution of the problem.

As the major part of this work is not the modelling of our setting, but the numerical implementation and the comparison of different models, we refer to [Bet93, Hau02, WBH08] for details on the continuum mechanical point of view on the modelling and on thermodynamical consistency.

During this whole thesis we consider our steel specimen to be the closure of a Lipschitz-bounded domain $\Omega \subset \mathbb{R}^3$. In the first four sections we assume all appearing quantities

to be as smooth as necessary. An exact definition will be given in the fifth section at the
weak formulation of the problem.

2.1   Heat equation

The modelling of heat conduction is known for very long time and can easily be found
in the standard literature. The heat equation can be deduced from a fundamental law of
thermodynamics: the *conservation of energy*. As we do not want to go into details here,
we present a brief and simplified approach of the modelling and refer to [Hau02, Wol08]
for a detailed explanation.

For the modelling of the heat equation we look at an arbitrary subset ω of Ω (with
regular boundary). For any time interval (t_1, t_2) the *amount of change of thermal energy*
is equal to the *flux of thermal energy* over the boundary of ω plus the thermal energy
which is released or consumed inside ω by inner *sources*, e.g. phase transitions or chemical
reactions.

The *amount of change of thermal energy* in ω is described by the first integral of equation (2.1.1). In the integrand there is the time derivative of the temperature θ multiplied
by the initial density ρ_0 and the specific heat c_e. The specific heat is temperature dependent, later in Section 2.4, where the coupled problem is treated, we will see, that it also
depends on the phases. For brevity we will not write down these dependencies explicitly.

Now we look at the *flux of thermal energy* over the surface of ω. According to Fourier's
Law this flux between two points is (locally) proportional to their temperature difference.
So we can define the heat flux vector $\boldsymbol{q}$ as

$$\boldsymbol{q} = -\kappa \nabla \theta$$

where κ is the heat conductivity of the material. The heat conductivity is temperature
dependent (it also depends on the phases, as we will see in Section 2.4, where the coupled
problem is treated). When we multiply the heat flux vector $\boldsymbol{q}$ by the outer normal vector
$\boldsymbol{n}$ and integrate over the surface of ω and the time interval, then we obtain the first term
on the right hand side of equation (2.1.1).

Finally we assume, that the inner *sources* in Ω have a density r and we denote by $\hat{r}$
its restriction to ω. Now we can model the amount of thermal energy which is released
or consumed inside ω by the last term in equation (2.1.1). Possible contributions to r
(in our setting) are: the thermal energy released or consumed by phase transition (latent

heat) or by mechanical deformation (dissipation).

$$(2.1.1) \qquad \int_{t_1}^{t_2} \int_\omega \rho_0 c_e \frac{d\theta}{dt} \, dx dt = \int_{t_1}^{t_2} \int_{\delta\omega} -\kappa \frac{\delta\theta}{\delta n} \, ds dt + \int_{t_1}^{t_2} \int_\omega \hat{r} \, dx dt$$

By applying Gauss' divergence theorem and some basic transformations we obtain the following equation:

$$(2.1.2) \qquad \int_{t_1}^{t_2} \int_\omega \left(\rho_0 c_e \frac{d\theta}{dt} - \operatorname{div}(\kappa\nabla\theta) - \hat{r} \right) dx dt = 0$$

Since ω as well as (t_1, t_2) were chosen arbitrarily we can conclude that the integrand of (2.1.2) must be zero. This gives us a first formulation of the heat equation.

$$(2.1.3) \qquad \rho_0 c_e \frac{d\theta}{dt} - \operatorname{div}(\kappa\nabla\theta) = r \quad \text{in } \Omega \times (0, T)$$

Remark. The above equation is a partial differential equation of parabolic type. The dependence of the specific heat, c_e, and the heat conductivity, κ, on the temperature make it a non-linear equation.

To complete this problem we add initial conditions

$$(2.1.4) \qquad \theta(\boldsymbol{x},0) \equiv \theta_0 \quad \text{in } \Omega$$

and conditions on the boundary values. As it fits best in our problem setting we choose a Robin boundary condition. Of course, Dirichlet and Neumann boundary conditions are possible, too. In our case the heat flux over the boundary is proportional to the temperature difference between the interior and the external temperature θ_{ext}. We obtain the following equation, where δ is the heat transfer coefficient.

$$(2.1.5) \qquad -\kappa\nabla\theta \cdot n = \delta(\theta - \theta_{ext}) \quad \text{on } \partial\Omega \times (0, T)$$

In this way equations (2.1.3), (2.1.4) and (2.1.5) give us the complete initial boundary value problem for heat conduction.

Remark. The boundary condition (2.1.5) describes the convection process, which is dominant in our case. The additionally occurring radiation is neglected here, but it could be integrated via an additional fourth-order term $\epsilon K_B(\theta^4 - \theta_{ext}^4)$, where K_B is the Boltzmann constant and ϵ the emissivity.

2.2 Phase transition

As already mentioned in the remarks on steel, Section 1.1, steel is a polycrystalline material consisting mainly of iron and carbon. At certain temperatures the crystal lattices become unstable and solid-solid phase transformations take place. While for iron this change is relatively easily described, the carbon atoms in steel complicate this process and the topic is still subject to intensive research. The modelling of this phenomenon is advanced on all levels: micro-, meso- and macroscopic. In this section we first want to give an idea of the microscopic changes which steel undergoes when quenched. This is important for the understanding of the need of different models for different types of phase transformations. Subsequently we will deal with macroscopic and phenomenological models for the different phases.

Concerning the quenching of steel one distinguishes generally between two different classes of phase transition: the diffusion-controlled and the diffusion-less transformations. In the *diffusion-controlled transformation,* carbon atoms have time to diffuse from the face-centered cubic austenite lattice into the body-centered cubic lattice of the forming phase, for 100Cr6 these phases are: pearlite[1] and bainite (and cementite, which is not considered in this work), compare Figure 1.3 for the different lattices. The *diffusion-less transformation* takes place instantly when the austenite lattice changes to the body-centered tetragonal lattice of the forming phase: martensite.

In this work we will not discuss micro- or mesoscopic models, but concentrate on macroscopic, phenomenological models for phases which appear in the quenching of 100Cr6 steel, starting with austenite only. Before presenting the models themselves we have to formulate some important assumptions. Since we are dealing only with the macroscopic situation, we assume that in every point $\boldsymbol{x}$ of our domain all m phases are given in phase fractions $p_i(\boldsymbol{x})$ which sum up to one.

$$\sum_{i=1}^{m} p_i(t, \boldsymbol{x}) = 1 \qquad t \in [0, T], \, \boldsymbol{x} \in \Omega$$

We define a vector of phases P,

$$(P)_i := p_i, \quad i = 1, \dots, m,$$

whose first component, p_1, is always set to be the phase fraction of austenite. This is

---

[1] Although pearlite is a two-phased lamellar structure, which consists of ferrite and cementite, we will refer to it as phase.

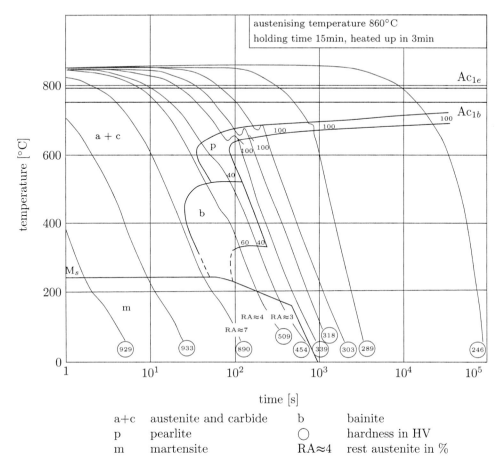

Figure 2.1: Time-temperature-transformation (TTT) diagram of 100Cr6 bearing steel, diagram taken from [Max61].

convenient, because in our applications only austenite transforms into other phases.

Furthermore the transformations of different phases do not interact. Each phase has a temperature interval in which a transformation is possible, compare Figure 2.1 where a time-temperature-transformation diagram of 100Cr6 steel is shown. For a phase fraction p_i we define the maximal possible phase fraction for the current temperature by $\bar{p}_i(\theta)$, for diffusion-controlled transformations $\bar{p}_i(\theta)$ is often called equilibrium phase fraction.

For simplicity we will formulate the models for a single phase p forming from austenite. The case where multiple phases are forming can be found in [WBB07].

To model the diffusion-controlled phase transformations we use the widespread *Johnson-Mehl-Avrami* model, which is written here in form of an ordinary differential equation.

$$(2.2.1) \qquad \dot{p} = -(p - \bar{p})\frac{n}{\tau} \left(-\ln\left(1 - \frac{p}{\bar{p}}\right)\right)^{1 - \frac{1}{n}}$$

Here $\bar{p}$ is the equilibrium phase fraction and n, τ are temperature dependent material parameters. The forming of the diffusion-controlled phases pearlite, bainite and cementite can be described by this model using different parameters $n, \tau, \bar{p}$. This model works particularly well for the description of isothermal transformations, but is also used for non-isothermal transformations, compare e.g. [BHSW04].

In [WBBL07] Wolff et. al. compared several modifications of the Johnson-Mehl-Avrami equations with experimental data. As there was no modified model witch worked best for all situations, we keep at the classical form.

For the diffusion-less martensite transformation we use, amongst others, the well-known *Koistinen-Marburger* model [KM59]

$$(2.2.2) \qquad p(\theta(t)) = 1 - \exp^{-\frac{\theta_{ms} - \theta(t)}{\theta_{m0}}}$$

where θ_{ms} is the martensite start temperature and θ_{m0} a material parameter. The Koistinen-Marburger model can easily be obtained as ordinary differential equation from the above equation.

In Section 4.2 we compare also two other models for martensite forming:

$$(2.2.3) \qquad \dot{p}(t) = \max\left(0, (\bar{p}(\theta(t)) - p(t))\mu\right)$$

the model of *Leblond-Devaux* [LD84] (where $\bar{p}(t)$ is the at time t and temperature θ maximal possible phase fraction given by the Koistinen-Marburger equation)
and the model of *Yu* [Yu97]:

$$(2.2.4) \qquad p(\theta(t)) = \frac{\theta_{ms} - \theta(t)}{\theta_{ms} - \theta(t) - \beta(\theta_{mf} - \theta(t))}$$

(where θ_{mf} is the martensite end temperature and β is a material parameter).

A comparison between different models is necessary, as there exist several models for martensitic transformation and it is not a priori clear which model works best.

An effect not considered in this work is the so-called stress- or strain-dependent phase transformation. If. e.g. the austenite is plastically deformed before the phase transformation takes place without applied stress, then the phase transformation is accelerated. This effect is stronger for a plastic deformation in uni axial tension than in uni axial compression, compare [Ahr03] and references therein. New experimental results within the SFB 570 indicate, that this effect is negligible for 100Cr6 steel.

2.3   Deformation behaviour

This section is devoted to the modelling of the deformation behaviour. In the first Subsection 2.3.1 we will deduce the basic deformation equation via the law of balance of momentum, which is fundamental in continuum mechanics. In the following three subsections we will introduce three different types of material behaviour. At first, in Subsection 2.3.2 elasticity will be treated, followed by classical plasticity in Subsection 2.3.3 and transformation induced plasticity (TRIP) in Subsection 2.3.4. The model of classical plasticity which we will introduce is independent of the velocity of applied forces which is the main difference to another class of material behaviour: viscoplasticity.

In this work we are considering infinitesimal strain which means that the occurring deformations are "small". This is a major assumption that simplifies certain parts of the modelling. The assumption of small strains is valid for the settings of heat treatment with moderate external forces, we are dealing with. Here we give only a short overview of the modelling, see for example [BG02, Bet93, FLS91, HR99, Hau02] for a more detailed explanation.

The first work on *elasticity theory* was done by Hook in 1678 when he discovered that the elongation is proportional to the applied load. Compared to this, the *theory on plasticity* is relatively young. In 1864 Tresca formulated his yield condition, followed by De Saint Venant and Levy in 1870 with first studies of constitutive equations. Von Mises formulated a second yield condition in 1913 and there was progress in the mathematical theory about the middle of the 20th century, while the application to real problems had to wait until 1970 when computers became available and sufficiently fast. The effect of *transformation induced plasticity* is even younger, first important work was done by Greenwood and Johnson and by Magee in the 1960's.

2.3.1   Balance of momentum

In this section we introduce the balance of momentum as it is the basis for elasticity, plasticity and transformation induced plasticity which are dealt with in the following subsections.

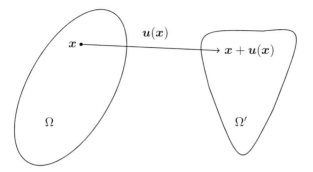

Figure 2.2: Reference and deformed configuration

We define Ω to be our reference configuration and $\boldsymbol{u} : \bar{\Omega} \times [0, T] \rightarrow \mathbb{R}^3$ to be the displacement vector, see Figure 2.2.

In the next step we define the strain tensor $\boldsymbol{\epsilon} : \bar{\Omega} \times [0, T] \rightarrow \mathbb{R}^3$ as follows:

$$\boldsymbol{\epsilon} := \frac{1}{2} \left(\nabla \boldsymbol{u} + (\nabla \boldsymbol{u})^T \right)$$

The form of the strain tensor is special, because we are considering small strains only. If we considered finite strain, the strain tensor would have an additional term $\frac{1}{2} \nabla \boldsymbol{u} (\nabla \boldsymbol{u})^T$, which is neglected for small strains. The diagonal elements of the strain tensor are called direct strains. When we imagine a fiber parallel to an axis attached at an arbitrary point of the reference configuration, then the respective diagonal element of the strain tensor equals half the length change of the fibre in the deformed configuration along the respective axis. The off-diagonal elements are referred to as shear strains. We choose two fibres of equal length to lie parallel to two axes, then the respective off diagonal element gives us a measure of the change in angle between those two fibres in the deformed configuration.

As a next step we will introduce Cauchy's axiom and explain the Cauchy stress vector and the Cauchy stress tensor. May Ω be in an deformed configuration and let $\partial \omega$ be an oriented surface on Ω, which has at the point $\boldsymbol{x}$ the normal $\boldsymbol{n}$. We look at the force applied by the material on the side to which $\boldsymbol{n}$ is pointing upon the material on the other side. Cauchy's axiom says, that this force (per unit area) depends on $\partial \omega$ only through $\boldsymbol{n}$ and is called the Cauchy stress vector $\boldsymbol{t}(\boldsymbol{n})$. When $\partial \omega$ belongs to the boundary of Ω, then $\boldsymbol{t}(\boldsymbol{n})$ represents the contact force applied on Ω by the surrounding environment. The Cauchy stress tensor, $\boldsymbol{\sigma} : \bar{\Omega} \times [0, T] \rightarrow \mathbb{R}^3$, which is a second order tensor, is introduced by the following correlation:

(2.3.1) $$t(\boldsymbol{n}) = \boldsymbol{\sigma} \cdot \boldsymbol{n}$$

(2.3.2) $$\boldsymbol{\sigma} = \boldsymbol{\sigma}^T$$

To derive the constitutive equation, we proceed analogously as in the case of the heat equation. We choose an arbitrary (regularly bounded) subset ω of Ω. The balance of linear and angular momentum yields that the sum of all external forces is equal to the sum of all inertia forces. The external forces consist of external surface forces $\boldsymbol{\sigma} \cdot \boldsymbol{n}$ (first term of equation (2.3.3)) and body forces $\hat{\boldsymbol{f}}$ (second term of equation (2.3.3)), e.g. gravity. The acceleration is given by $\frac{\partial^2 \boldsymbol{u}}{\partial t^2}$, thus the inertia forces are given by the third term of equation (2.3.3).

(2.3.3) $$\int_{\partial \omega} \boldsymbol{\sigma} \cdot \boldsymbol{n} ds + \int_{\omega} \rho \hat{\boldsymbol{f}} dx = \int_{\omega} \rho \frac{\partial^2 \boldsymbol{u}}{\partial t^2} dx$$

Again we use Gauss' divergence theorem and obtain:

(2.3.4) $$\int_{\omega} \left(\operatorname{div}(\boldsymbol{\sigma}) + \rho \hat{\boldsymbol{f}} - \rho \frac{\partial^2 \boldsymbol{u}}{\partial t^2} \right) dx = 0$$

As ω was chosen arbitrarily, the integrand of (2.3.4) has to be zero. With this result we can formulate the hyperbolic deformation equation as initial boundary value problem.

(2.3.5a) $$\rho \frac{\partial^2 \boldsymbol{u}}{\partial t^2} - \operatorname{div}(\boldsymbol{\sigma}) = \rho \boldsymbol{f}$$ in $\Omega \times [0, T]$

(2.3.5b) $$\boldsymbol{u}(0) = 0 \text{ and } \frac{\partial \boldsymbol{u}}{\partial t}(0) = 0$$ in Ω

(2.3.5c) $$\boldsymbol{u} = g_D$$ on $\Gamma_D \times [0, T]$

(2.3.5d) $$\boldsymbol{\sigma} \cdot \boldsymbol{n} = g_N$$ on $\Gamma_N \times [0, T]$

Here we chose Dirichlet boundary conditions to demand the displacement g_D on Γ_D and Neumann boundary conditions to apply an external force g_N on Γ_N, where $\partial \Omega = \Gamma_D \cup \Gamma_N$. Please note that homogeneous initial conditions were chosen to model quenching, for other settings it might be necessary to choose different initial conditions.

Remark. In the above form equation (2.3.5a) is a partial differential equation of hyperbolic type. In the applications we are dealing with, the inertia term is often neglected, compare Section 4.3. In this case the deformation equation is of elliptic type.

Equation (2.3.5a) is derived directly from a fundamental law of continuum mechanics and is the basis for our further proceeding. At the moment problem (2.3.5) is under-determined, because there is no relation specified between $\boldsymbol{\sigma}$ and $\boldsymbol{\epsilon}(\boldsymbol{u})$. This relation will be formulated in the following sections for three different types of material behaviour: elasticity, plasticity and transformation induced plasticity. Together with the balance of momentum this will complete the mathematical model for the mechanical behaviour of steel.

2.3.2   Elasticity

Elasticity is the simplest of the three types of material behaviour considered. A body is said to behave elastically if there is a functional correlation between stress and strain. This means, that it deforms under external or internal forces, but returns (on the same load path) to its original shape if the forces are removed.

In *linear* elasticity we have the correlation between the strain and the stress tensor given by the well-known Hooke's law:

$$(2.3.6) \qquad\qquad\qquad \boldsymbol{\sigma} = \boldsymbol{C}\boldsymbol{\epsilon}$$

where $\boldsymbol{C}$ is the elasticity tensor, a linear map from the space of symmetric matrices into itself, and may be represented as a fourth order tensor with the following symmetry properties:

$$C_{ijkl} = C_{jikl} = C_{ijlk} = C_{klij}$$

In this work we are dealing with isotropic materials which are an important group of materials. Isotropy in the elastic behaviour means that the response to a force is inde-pendent of the direction of the force. This property reduces the twenty-one independent elements of the elasticity tensor to two. There are three pairs of elastic quantities, which are wide spread for the description of the elasticity tensor (see next remark). We use the commonly spread Lamé coefficients λ, μ. Then the elasticity tensor has the following form

$$C_{ijkl} = \lambda \delta_{ij}\delta_{kl} + \mu(\delta_{ik}\delta_{jl} + \delta_{il}\delta_{jk})$$

and the stress tensor is given by:

$$(2.3.7) \qquad\qquad\qquad \boldsymbol{\sigma} = \lambda(\operatorname{tr}(\boldsymbol{\epsilon}))\boldsymbol{I} + 2\mu\boldsymbol{\epsilon}$$

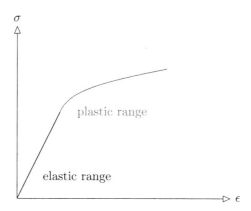

Figure 2.3: Stress-strain diagram of a tensile test of a material showing first linear elasticity and then plasticity with hardening.

Remark. The Lamé coefficients are connected to two other important elastic quantities: the Young modulus E and the Poisson ratio ν. The Young modulus corresponds to the slope of the stress-strain curve in its elastic range (see Figure 2.3) and is comparable to the stiffness of a spring. The Poisson ratio measures the lateral contraction of the material. In engineering literature one as well finds the bulk modulus K and the shear modulus G. These quantities share the following relations:

$$\lambda = \frac{\nu E}{(1+\nu)(1-2\nu)} \qquad \mu = \frac{E}{2(1+\nu)} \qquad K = \frac{E}{3-6\nu}$$

$$E = \frac{\lambda}{2(\lambda+\mu)} \qquad \nu = \frac{\mu(3\lambda+2\mu)}{\lambda+\mu} \qquad G = \frac{E}{2+2\nu}$$

Remark. Later on, when we consider the coupled problem, the Lamé coefficients will depend on both temperature and phases.

Remark. As we are considering small deformations in steel, it is sufficient to consider *linear elasticity*. For large deformations of other materials, e.g. rubber, it would be necessary to deal with a *non linear elasticity* model.

Bringing together problem (2.3.5) and equation (2.3.7), we obtain the constitutive equation of linear elastic deformation behaviour, which can be rewritten to the following form, compare [Bet93].

(2.3.8a) $\mu \Delta \boldsymbol{u} + (\mu + \lambda) \operatorname{grad} \operatorname{div} \boldsymbol{u} + \rho \boldsymbol{f} = 0$ in Ω

(2.3.8b) $\boldsymbol{u} = g_D$ on $\Gamma_D \times [0, T]$

(2.3.8c) $\left(\lambda (\operatorname{tr}(\boldsymbol{\epsilon}(\boldsymbol{u}))) \boldsymbol{I} + 2\mu \boldsymbol{\epsilon}(\boldsymbol{u}) \right) \cdot \boldsymbol{n} = g_N$ on $\Gamma_N \times [0, T]$

Now we can go on to add more complicated deformation models to our problem setting.

2.3.3   Plasticity

Phenomenology

To introduce plastic deformation behaviour we consider a one-dimensional tensile test of a cylinder specimen. While the applied load is within the elastic range, there will be no permanent deformation after unloading. When we increase the load (and hence the stress in the specimen) beyond a material parameter called yield stress R_0, then many materials will show plastic flow, so that after unloading a permanent strain ϵ^{cp} is left. There are different models which describe this plastic flow. Here we deal with perfect plasticity, plasticity with isotropic hardening and plasticity with isotropic and kinematic hardening.

   To give another descriptive explanation of classical plasticity we recall that metals are polycrystalline materials. An elastic deformation corresponds to a variation in the inter-atomic distances, while a plastic deformation means that there are slip movements with changes of inter-atomic bonds.

   When coming back to our one-dimensional tensile test, we see in Figure 2.4(a) the stress-strain diagram of perfect plasticity. In this model the stress remains constant during the flow. This is the case for mild steel whose stress-strain diagram has a plateau zone after the elastic area. For the steels we are dealing with, this model is not applicable.

   Another important model is plasticity with isotropic hardening. Hardening means that the plastic flow only occurs if the stress increases and that the yield stress increases during the flow from R_0 to $R_0 + R$, where R_0 is the initial yield stress and R is the increase of the yield stress. So when the specimen is unloaded and reloaded, it will not yield until the stress exceeds the stress before unloading $(R_0 + R)$, comp. Figure 2.4(b). (As we are dealing with hardening only and do not treat softening, $\dot{R} \geq 0$ holds for all times.) A descriptive explanation of the hardening effect on the atomic level is the following. During the plastic flow, metals show an increase of dislocations[2] which propagate through the

---

[2]Compare Figure 1.4 for a graphic representation of lattice imperfections.

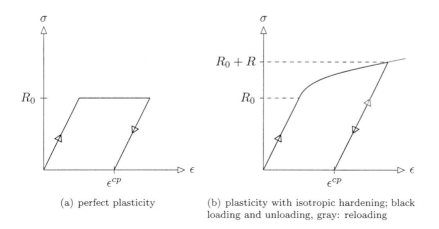

(a) perfect plasticity (b) plasticity with isotropic hardening; black loading and unloading, gray: reloading

Figure 2.4: 1d-tensile test, stress-strain diagram with different types of hardening

lattice along slip planes. The dislocations pile up at grain boundaries or other dislocations or one dislocation must cut through another one, needing more energy to do so. In this way the dislocations tend to block each other, so that at the reloading the steel shows more resistance till it yields. Compare [SD96] for a detailed explanation of plasticity in metals.

By convention tensile stresses have a positive sign and compression stresses are negative. In the models of perfect plasticity and of plasticity with isotropic hardening the elastic behaviour is symmetric for tension and compression. Their elastic ranges, $\mathbb{E}$, are $[-R_0, R_0]$ and $[-(R_0 + R), (R_0 + R)]$ respectively. In Figure 2.6(a) for the case of plasticity with isotropic hardening the specimen was first loaded in tension and subsequently loaded in compression.

In the model of plasticity kinematic hardening the center of the elastic range moves by the development of the back-stress X^{cp} thus the elastic range becomes $[-(R_0 - X^{cp}), (R_0 + X^{cp})]$, comp. Figure 2.6(b). This effect can only be understood on the atomic level. As already mentioned, during the plastic flow dislocations are generated and pile up at grain boundaries or other dislocations. This pileup produces a back stress opposite to the applied stress. When the applied stress is reduced or removed the deformation caused by the stress decreases. So the net force producing a plastic deformation is the difference between the applied stress and the back stress; see again [SD96] for details.

Experiments show that many metals show a combination of both described hardening behaviours. In the case of combined isotropic and kinematic hardening the elastic range both grows and moves: $[-(R_0 - X^{cp} + R), (R_0 + X^{cp}) - R]$, compare Figure 2.6(c).

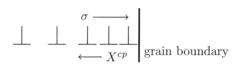

Figure 2.5: Pileup of dislocations at a barrier

Modelling plasticity in 1d

After dealing with the phenomenological aspects of classical plasticity we need to make a basic assumption to formulate the mathematical model. As already mentioned in the section of elastic behaviour we are considering small strains and isotropic materials. Additionally we assume that the total strain ϵ can be decomposed into an elastic and a plastic part:

$$\epsilon = \epsilon^{el} + \epsilon^{cp}$$

The model we will formulate in this section will be *rate independent*. For our considered material and processes we can neglect the dependence of the material on the rate at which an external force is applied. In contrast to this, viscoplasticity includes a dependence on the rate at which external forces are applied, compare e.g. [Hau02].

To derive a first mathematical model for plasticity with isotropic hardening, we are still looking at our one-dimensional tensile/compression test. We assume, that the elastic response of our specimen is given by:

$$(2.3.9) \qquad\qquad \sigma = E\epsilon^{el} = E(\epsilon - \epsilon^{cp})$$

To check whether for the actual stress in the specimen the response is elastic or plastic we define a yield-function $F(\sigma, R)$, with $F(\sigma, R) \leq 0$ for all physically admissible (σ, R). The material deforms plastically when the yield criterion is satisfied:

$$(2.3.10) \qquad\qquad F(\sigma, R) := |\sigma| - (R_0 + R) = 0$$

For $F(\sigma, R) < 0$ we have elastic behaviour, which means that there is no change in ϵ^{cp}. Moreover stresses which lead to $F(\sigma, R) > 0$ are not allowed, so changes in ϵ^{cp} can only happen while $F(\sigma, R) = 0$. The direction in which our specimen yields is equal to the direction of the applied stress, thus positive for tensile forces and negative for compression forces. This can be formulated by the plastic flow rule:

$$(2.3.11) \qquad\qquad \dot{\epsilon}^{cp} = \gamma \operatorname{sign}(\sigma) \Leftrightarrow F(\sigma, R) = |\sigma| - (R_0 + R) = 0$$

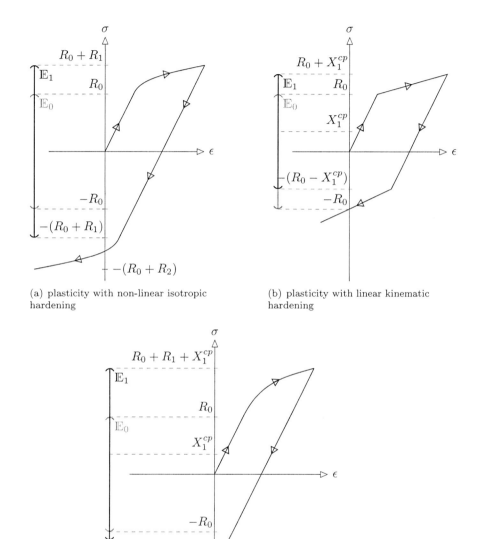

(a) plasticity with non-linear isotropic hardening

(b) plasticity with linear kinematic hardening

(c) plasticity with combined isotropic and kinematic hardening

Figure 2.6: 1d-tensile and compression test, stress-strain diagram with different types of hardening

where $\gamma \geq 0$ is the plastic multiplier, in this setting the absolute value of the slip rate. It can be shown, that the rate independence of the model follows from the above flow rule.

To clarify the interaction between elastic and plastic behaviour, we conclude that in the elastic range, the following holds:

$$F(\sigma, R) < 0 \Rightarrow \gamma = 0$$

while at plastic deformation it holds:

$$\gamma > 0 \Rightarrow F(\sigma, R) = 0$$

From these observations we derive the so called Kuhn-Tucker condition

(2.3.12) $$\gamma F(\sigma, R) = 0$$

Additionally we introduce an equation which is referred to as consistency condition, when we consider the case where $F(\sigma(t), R(t)) = 0$. We know that $\gamma > 0$ and can deduce that $\frac{d}{dt}F(\sigma(t), R(t)) \leq 0$ holds. Otherwise there would exist an Δt with $F(\sigma(t + \Delta t), R(t + \Delta t)) > 0$ which is not allowed. Therefore it is now demanded:

$$\gamma > 0 \qquad \Rightarrow \qquad \frac{d}{dt}F(\sigma(t), R(t)) = 0$$

$$\frac{d}{dt}F(\sigma(t), R(t)) < 0 \qquad \Rightarrow \qquad \gamma = 0$$

Which gives us the consistency condition:

(2.3.13) $$\gamma \frac{d}{dt}F(\sigma, R) = 0$$

To complete our model a specification of the hardening behaviour is necessary. There are different modelling approaches for isotropic as well as kinematic hardening. We will start with the relatively simple case of (exclusively) isotropic hardening for which we use the well-known Ramberg-Osgood model [RO43]. This hardening law involves material parameters c, m and the accumulated plastic strain s^{cp} and describes the development of the increase of the yield surface, R. We specify the Ramberg-Osgood model in a modified form:

(2.3.14) $$\dot{s}^{cp} := |\dot{\epsilon}^{cp}| = \gamma$$
(2.3.15) $$R = c\,(s^{cp})^m$$

For convenience we summarise:

2.3.1. 1d model for plasticity with isotropic hardening:

- *Hooke's law with additive decomposition of the strain tensor*

$$\sigma = E(\epsilon - \epsilon^{cp})$$

- *yield criterion*

$$F(\sigma, R) = |\sigma| - (R_0 + R) = 0$$

- *plastic flow rule*

$$\dot{\epsilon}^{cp} = \gamma \operatorname{sign}(\sigma)$$

- *hardening law: Ramberg-Osgood model*

$$\dot{s}^{cp} = \gamma \qquad R = c\left(s^{cp}\right)^m$$

- *Kuhn-Tucker and consistency condition*

$$\gamma F(\sigma, R) = 0 \qquad \gamma \frac{d}{dt} F(\sigma, R) = 0 \; (\textit{if } F(\sigma, R) = 0)$$

As a next step we add kinematic hardening behaviour. For this purpose we introduce the back-stress X^{cp} and modify the yield criterion F and the plastic flow rule. There are different modelling approaches for the development of the back-stress X^{cp}, in the first place we choose the wide-spread Prager model [Pra49, Pra56], which imposes a linear dependence on ϵ^{cp}.

2.3.2. modifications for isotropic and kinematic hardening:

- *yield criterion*

$$F(\sigma, R, X^{cp}) = |\sigma - X^{cp}| - (R_0 + R) = 0$$

- *plastic flow rule*

$$\dot{\epsilon}^{cp} = \gamma \, \text{sign}(\sigma - X^{cp})$$

- *hardening laws: Ramberg-Osgood and Prager model*

$$\dot{s}^{cp} = \gamma \qquad R = c\,(s^{cp})^m$$

$$\dot{X}^{cp} = c_{cp}\dot{\epsilon}^{cp}, \quad c_{cp} = const$$

Remark. Later on, when we consider the fully coupled problem, R, R_0, c, m and c_{cp} will depend on both temperature and phases.

For some materials the description of the hardening behaviour via the Ramberg-Osgood and Prager model is not good. Therefore we also introduce the Armstrong-Frederick model [AF07] for isotropic and kinematic hardening. In difference to the Ramberg-Osgood and Prager model, the back-stress evolves non-linearly. The additional term in the development of the back-stress is often referred to as recovery term and realises a saturation. For a review on the modelling approaches for hardening laws for other possible models see [Cha08].

2.3.3. Armstrong-Frederick model for isotropic and kinematic hardening:

- *hardening laws:*

$$\dot{R} = \hat{\gamma}\dot{s}^{cp} - \hat{\beta}R\dot{s}^{cp}$$
$$\dot{X}^{cp} = \hat{c}\dot{\epsilon}^{cp} - \hat{b}X^{cp}\dot{s}^{cp}$$

$$\hat{\gamma},\hat{\beta},\hat{c},\hat{b} = const$$

Remark. All four parameters in the Armstrong-Frederick model can be temperature dependent. When $\hat{\gamma}$ or $\hat{c}$ depend on the temperature there will be an additional term in the respective equation. We will discuss this further in Section 2.4, where the coupled problem is treated.

Remark. We can define the set of admissible stresses as follows:

(2.3.16) $$\mathbb{E} := \{(\sigma, R, X^{cp}) \in (\mathbb{R}, \mathbb{R}^+, \mathbb{R}) \mid F(\sigma, R, X^{cp}) \le 0\}$$

The interior of $\mathbb{E}$ is called the elastic range while its boundary is referred to as yield surface. It is obvious that for plasticity with hardening the material's behaviour is dependent on the load history.

Using our constitutive equations for plasticity, especially the Kuhn-Tucker and the consistency condition, we can distinguish four different loading cases:

$$F < 0 \Leftrightarrow (\sigma, R, X^{cp}) \in \text{int}(\mathbb{E}) : \qquad\qquad \text{elastic loading}$$

$$F = 0 \Leftrightarrow (\sigma, R, X^{cp}) \in \partial(\mathbb{E}) :$$

$$\frac{d}{dt}F < 0 \Rightarrow \gamma = 0 \qquad\qquad \text{elastic unloading}$$

$$\frac{d}{dt}F = 0 \text{ and } \gamma = 0 \qquad\qquad \text{neutral loading}$$

$$\frac{d}{dt}F = 0 \text{ and } \gamma > 0 \qquad\qquad \text{plastic loading}$$

Three-dimensional model

When shifting our model in Box 2.3.1 to three dimensions we have to be aware that certain quantities which were scalar before are now tensors, e.g σ, ϵ^{cp} and X^{cp}.

There exists an additive split of the strain tensor into a purely volumetric and a volume preserving part:

$$\epsilon = \epsilon^* + \frac{1}{3}\text{tr}(\epsilon)I$$

where the deviator of the strain tensor ϵ^* is volume preserving and $\frac{1}{3}\text{tr}(\epsilon)I$ is purely volumetric. There is an analogue split of the tensor ϵ^{cp}.

Since we want to apply our plasticity model to steel it is efficient to neglect the volumetric changes caused by classical plasticity because they are very small. It was very often observed for many material types (except porous media), that hydrostatic pressure and the corresponding volumetric changes are purely elastic. From this so-called *plastic in-compressibility assumption* we obtain:

$$\text{tr}(\epsilon^{cp}) = 0$$

For the formulation of a three-dimensional plasticity model we need at first a yield criterion. In the literature there are several approaches, but for the application to steel only two yield criterions are commonly used: the *von Mises-* and the *Tresca yield criterion.*

The *von Mises* yield criterion reads as follows:

$$(2.3.17) \qquad\qquad F = ||\boldsymbol{\sigma}^*|| - \sqrt{\frac{2}{3}}\left(R_0 + R\right) = 0$$

For an uni-axial stress state this criterion implies, that the yielding starts at $(R_0 + R)$. The scaling factor $\sqrt{\frac{2}{3}}$ arises because in standard norm for a one dimensional stress state with applied stress S it holds: $||\boldsymbol{\sigma}^*|| := \sqrt{\boldsymbol{\sigma}^* : \boldsymbol{\sigma}^*} = \sqrt{\frac{2}{3}}S$. Of course one could also use a non-standard scaled norm $||\boldsymbol{\sigma}^*||_s := \sqrt{\frac{3}{2}}||\boldsymbol{\sigma}^*||$ but this could easily cause confusion.

Tresca's yield criterion uses the eigenvalues of the stress tensor: $\sigma_1, \sigma_2, \sigma_3$. We define $\sigma_{max} = \max(\sigma_1, \sigma_2, \sigma_3)$ and $\sigma_{min} = \min(\sigma_1, \sigma_2, \sigma_3)$ to formulate the criterion

$$(2.3.18) \qquad\qquad F = (\sigma_{max} - \sigma_{min}) - \sqrt{\frac{2}{3}}\left(R_0 + R\right) = 0$$

Both criteria have the property of pressure-insensitivity, meaning that the hydrostatic pressure component of the stress tensor, $\frac{1}{3}\operatorname{tr}(\boldsymbol{\sigma})$, has *no* influence on the plastic yielding. For any pressure p it holds for both criteria:

$$F(\boldsymbol{\sigma} + p\boldsymbol{I}) = F(\boldsymbol{\sigma})$$

This property is essential for the modelling of metals. Other common yield criteria like the one of Mohr-Coulomb or the one of Drucker-Prager are pressure sensitive, they are often used for the modelling of soil, rocks or concrete.

A illustrative representation of von Mises and Tresca's yield surfaces can be seen in Figures 2.7(a) and 2.7(b).

In [HPS92] it is pointed out that Tresca's criterion incorrectly predicts the amount of work done in a deformation. It is said, that several experiments showed that the von Mises criterion represents in general the yielding behaviour of most ductile materials. In contrast to this, [dSNPO08] writes, that for many metals experimentally determined yield surfaces fall between the von Mises and Tresca surfaces. It does not seem to be clear a priori, which yield criterion should be used for a certain metal. As using Tresca's criterion would cause problems in formulating an numerical algorithm, we will use from now on

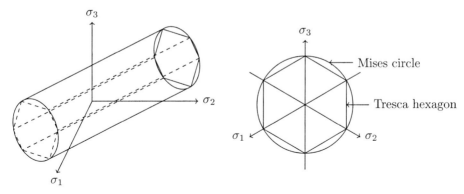

(a) yield surface in principal stress
(eigenvalues of $\boldsymbol{\sigma}$) space.

(b) yield criterion on the so-called π-plane:
$\sigma_1 + \sigma_2 + \sigma_3 = 0$

Figure 2.7: Comparison of von Mises and Tresca's yield criterion.

the von Mises yield criterion only.

After the formulation of the yield criterion we will introduce a plastic flow rule and a hardening law for the three dimensional isotropic case. For isotropic hardening we obtain the *plastic flow rule* by

$$(2.3.19) \qquad \dot{\boldsymbol{\epsilon}}^{cp} = \gamma \frac{\boldsymbol{\sigma}^*}{||\boldsymbol{\sigma}^*||}$$

In the most general form of a *hardening law* a set of internal hardening variables $\{\boldsymbol{\alpha}\}$ and hardening thermodynamical forces $\{\boldsymbol{A}\}$ are necessary. For the description of isotropic hardening of metals the set of internal hardening variables reduces to one scalar variable which determines the size of the yield surface. Here we will deal with strain hardening, so that this variable must be a measure for the strain: the accumulated plastic strain s^{cp}. It is defined as follows (see e.g. [dSNPO08, SH97] for details):

$$(2.3.20) \qquad s^{cp}(t) := \int_0^t \sqrt{\frac{2}{3}} ||\dot{\boldsymbol{\epsilon}}^{cp}(\tau)|| d\tau$$

The above expression can be simplified using the plastic flow rule (2.3.19):

$$(2.3.21) \qquad \dot{s}^{cp} = \sqrt{\frac{2}{3}} \gamma$$

The increase of the yield surface, R, is a function of the accumulated plastic strain.

Remark. In metal plasticity one often uses a so-called *associative plasticity model*, where both plastic flow rule and a hardening law are determined by derivatives of the yield function. For this relatively simple material behaviour equations (2.3.19) and (2.3.21) remain unchanged, so for this work it makes no difference, which way of modelling is chosen. Both approaches have advantages as well as disadvantages, which reveal, for example, when modelling more complex material behaviour, see [Wol08] for a detailed discussion.

We can now summarise the three dimensional plasticity model with isotropic hardening after the Ramberg-Osgood model.

2.3.4. 3d model for plasticity with isotropic hardening:

- *representation of the stress tensor*

$$\boldsymbol{\sigma} = 2\mu(\boldsymbol{\epsilon} - \boldsymbol{\epsilon}^{cp}) + \lambda \operatorname{tr}(\boldsymbol{\epsilon})\boldsymbol{I}$$

- *yield criterion (von Mises)*

$$F = ||\boldsymbol{\sigma}^*|| - \sqrt{\frac{2}{3}}\left(R_0 + R\right) = 0$$

- *plastic flow rule*

$$\dot{\boldsymbol{\epsilon}}^{cp} = \gamma \frac{\boldsymbol{\sigma}^*}{||\boldsymbol{\sigma}^*||}$$

- *hardening law: Ramberg-Osgood model*

$$\dot{s}^{cp} = \sqrt{\frac{2}{3}}\gamma \qquad R = c(s^{cp})^m$$

- *Kuhn-Tucker and consistency condition*

$$\gamma F = 0 \qquad \gamma \frac{d}{dt}F = 0 \ (if\ F = 0)$$

Remark. The associative plasticity model with von Mises yield criterion is in engineering literature often referred to as Prandtl-Reuss plasticity.

Before we specify the modified equations for plasticity with isotropic and kinematic hardening using the Prager model, we note that the back stress tensor is traceless, as it is shown in [Wol08, WBMS].

$$\text{tr}(\boldsymbol{X}^{cp}) = 0$$

This leads to the following equations:

2.3.5. modifications for isotropic and kinematic hardening:

- *yield criterion*

$$F = ||\boldsymbol{\sigma}^* - \boldsymbol{X}^{cp}|| - \sqrt{\frac{2}{3}}\left(R_0 + R\right) = 0$$

- *plastic flow rule*

$$\dot{\boldsymbol{\epsilon}}^{cp} = \gamma \frac{\boldsymbol{\sigma}^* - \boldsymbol{X}^{cp}}{||\boldsymbol{\sigma}^* - \boldsymbol{X}^{cp}||}$$

- *hardening laws: Ramberg-Osgood and Prager model*

$$\dot{s}^{cp} = \sqrt{\frac{2}{3}}\gamma \qquad R = c(s^{cp})^m$$

$$\dot{\boldsymbol{X}}^{cp} = \frac{2}{3}c_{cp}\dot{\boldsymbol{\epsilon}}^{cp}, \quad c_{cp} = const$$

As in the one-dimensional case we also specify the Armstrong-Frederick model for combined isotropic and kinematic hardening.

2.3.6. Armstrong-Frederick model for isotropic and kinematic hardening:

- *hardening laws:*

$$\dot{R} = \hat{\gamma}\dot{s}^{cp} - \hat{\beta}R\dot{s}^{cp}$$

$$\dot{\boldsymbol{X}}^{cp} = \frac{2}{3}\hat{c}\dot{\boldsymbol{\epsilon}}^{cp} - \hat{b}\boldsymbol{X}^{cp}\dot{s}^{cp}$$

$$\hat{\gamma}, \hat{\beta}, \hat{c}, \hat{b} = const$$

Remark. Please note, that one has to take care when comparing the back stress of the one dimensional setting, X^{cp}, to the back stress in the three dimensional setting, $\boldsymbol{X}^{cp}$, for a one dimensional stress state. In the one dimensional setting a tension test with applied stress S, plastic strain ϵ^{cp} and back stress X^{cp} corresponds in the three dimensional case to

$$(2.3.22) \qquad \boldsymbol{\sigma} = \begin{pmatrix} S & 0 & 0 \\ 0 & 0 & 0 \\ 0 & 0 & 0 \end{pmatrix} \quad \boldsymbol{X}^{cp} = \begin{pmatrix} X^{cp} & 0 & 0 \\ 0 & -\frac{1}{2}X^{cp} & 0 \\ 0 & 0 & -\frac{1}{2}X^{cp} \end{pmatrix}$$

The one dimensional yield condition reads:

$$|S - X^{cp}| - (R_0 + R) = 0$$

while the three dimensional yield condition becomes the following equation when one inserts (2.3.22):

$$\|\boldsymbol{\sigma}^* - \boldsymbol{X}^{cp}\| - \sqrt{\frac{2}{3}}\left(R_0 + R\right) = 0$$

$$\sqrt{\frac{2}{3}}|S - X^{cp}| - \sqrt{\frac{2}{3}}(R_0 + R) = 0$$

One can easily see, that the following relation between the one and the three-dimensional back stress is valid:

$$X^{cp} = \frac{3}{2}\boldsymbol{X}^{cp}_{11}$$

Remark. The modelling of classical plasticity can also be done in a thermodynamic framework, see e.g. [dSNPO08, HR99, Hau02, Hel98].

2.3.4 Transformation induced plasticity

Transformation induced plasticity is an important factor of distortion of steel. Its main characteristic is the occurrence of macroscopic permanent deformations during phase transitions although the macroscopic yield stress of the weaker phase is not reached.

Although this phenomenon was worked on since the 1960's its modelling is still subject to intensive research. In many situations, as in our case, the main interest is directed towards a good reflection of the macroscopic material behaviour. That is why many models are purely phenomenological and do not rely on the mesoscopic effects transformation induced plasticity is caused by. There are modelling efforts on the meso- and

microscopic level, but for brevity we do not want to discuss these here and refer to
[Wol08, WBH08, WBS09] for a broader overview of modelling transformation induced
plasticity and references herein.

It is generally agreed that transformation induced plasticity is caused by two mesoscopic effects, the Greenwood-Johnson-effect and, in the case of martensitic phase transition, the Magee-effect.

The *Greenwood-Johnson-effect* was first mentioned in [GJ65] and states that differences in density between the existing and the forming phase cause local classic-plastic
and creep-deformations. When there are deviatoric stresses these local deformations lead
to macroscopic permanent strain.

The *Magee-effect* [Mag66] accounts for microscopic changes during martensite transformation. The atomic lattice of austenite shears spontaneously and forms in this way
martensite plates or needles whose orientation cause internal stresses. When the transformation is stress-free, these internal stresses neutralise each other due to the polycrystalline structure of steel. In the case of deviatoric stress the martensite plates form along
a preferred orientation and cause permanent deformations.

In macroscopic models for transformation induced plasticity the Greenwood-Johnson
and the Magee effect are often joined.

We will now describe in detail the model of transformation induced plasticity we are
working with and the model of Leblond which is also widespread. Both models are purely
phenomenological and refer to homogeneous materials. In [HFP+07] Hunkel et. al. investigated size changes of steel which is anisotropic due to segregation. It was shown in
[SSBH07] that the macroscopic behaviour of periodically layered steel materials can be
qualitatively recovered from mesoscopic simulations, where the model of transformation
induced plasticity was applied to a geometry small enough to dissolve the micro structure
of the material. For a quantitative description of the behaviour other models or methods
have to be applied.

To start with our model we define transformation induced plasticity as the *complete*
permanent strain occurring during phase transition while the yield stress is not reached.
This assumption differs from the Leblond model where the permanent strain is split into
a transformation induced and a classic-plastic part, we will come to this later.

At first we consider a situation where the yield stress is not reached. Then we use our
assumption to split the total strain into an thermoelastic part ϵ^{el} and an transformation
induced part ϵ^{up}.

$$(2.3.23) \qquad\qquad\qquad \epsilon = \epsilon^{el} + \epsilon^{up}$$

This enables us to deal with transformation induced plasticity and classical plasticity separately, in difference to the later discussed Leblond-model. In our model the volumetric changes caused by the phase transition is part of the thermoelastic strain. Therefore there are no volumetric changes caused by transformation induced plasticity:

$$(2.3.24) \qquad\qquad\qquad\qquad \mathrm{tr}(\boldsymbol{\epsilon^{up}}) = 0$$

Now we formulate our model for transformation induced plasticity for one phase forming, which is based on *Greenwood-Johnson* [GJ65] and purely phenomenological.

$$(2.3.25) \qquad\qquad\qquad \dot{\boldsymbol{\epsilon}}^{up} = \frac{3}{2}\kappa^{up}\boldsymbol{\sigma}^*\dot{\Phi}(p)\dot{p}, \quad \dot{p} \geq 0$$

Here κ^{up} is the Greenwood-Johnson parameter and Φ a monotone saturation function with $\Phi(0) = 0$ and $\Phi(1) = 1$ and $\Phi \in C[0,1] \cap C^1(0,1)$. In our approach the parameter κ^{up} has to be determined by experiments (unlike in the Leblond model) and appears to depend on the temperature and the current phase fraction. In [WBDH08] the Greenwood-Johnson parameter of 100Cr6 steel was considered to be stress dependent. This approach is no longer followed, as the experiments used in this article turned out to be insufficiently reliable.

There are several choices for Φ, the easiest one is the identity, but also others are widespread and the approach of Wolff was recently developed in [WBDH08].

$$(2.3.26) \qquad\qquad \Phi(p) = p \qquad\qquad\qquad\qquad\qquad \text{Tanaka}$$

$$(2.3.27) \qquad\qquad \Phi(p) = p(1 - \ln(p)) \qquad\qquad\qquad \text{Leblond}$$

$$(2.3.28) \qquad\qquad \Phi(p) = p(2 - p) \qquad\qquad\qquad\qquad \text{Denis/Desalos}$$

$$(2.3.29) \qquad\qquad \Phi(p) = \frac{1}{2}\left(1 + \sin\left(\pi p - \frac{\pi}{2}\right)\right) \qquad \text{Wolff}$$

In Section 4.2.3 we will present simulations with a direct comparison of different saturation functions, see also [SFHW09].

When we are dealing with more than one forming phase, in equation (2.3.25) the factor $\dot{\Phi}(p)\dot{p}$ (with $\dot{p} \geq 0$) changes to $\sum_{i=2}^m \dot{\Phi}_i(p_i)\dot{p}_i$, where the p_1 is the decreasing austenite. This is due to the fact that for the transformation induced plasticity only the actually growing phases, $\dot{p}_i \geq 0$, are considered, this also guarantees a non-negative factor in (2.3.25).

Furthermore we extend our model with a back-stress $\boldsymbol{X}^{up}$ which we define analogously to the back-stress for classical plasticity, compare [WBD$^+$07, WBH08].

$$(2.3.30) \qquad \dot{\boldsymbol{X}}^{up} = \frac{2}{3}c_{up}\dot{\boldsymbol{\epsilon}}^{up}, \qquad\qquad c_{up} = \text{const}$$

$$(2.3.31) \qquad \dot{\boldsymbol{\epsilon}}^{up} = \frac{3}{2}\kappa^{up}\left(\boldsymbol{\sigma}^* - \boldsymbol{X}^{up}\right)\dot{\Phi}(p)\dot{p}, \qquad\qquad \dot{p} \geq 0$$

The back-stress can also be coupled with classical plasticity, see Section 2.4.

Another well-known model for transformation induced plasticity and classical plasticity is the *Leblond* model. It was developed by Leblond and co-workers in [LMD86a, LMD86b, LMD89a, LMD89b]. The major difference to our model is the definition of transformation induced plasticity. Transformation induced plasticity strain is caused *only* by phase transitions while permanent deformations arising from changes of temperature or stress are understood as classical plasticity, even when the ultimate stress is not reached. In formulas this reads as follows:

$$(2.3.32) \qquad \boldsymbol{\epsilon} = \boldsymbol{\epsilon}^{el} + \boldsymbol{\epsilon}^p$$

$$(2.3.33) \qquad \dot{\boldsymbol{\epsilon}}^p = \underbrace{c_1\dot{\boldsymbol{\sigma}} + c_2\dot{\theta}}_{\dot{\boldsymbol{\epsilon}}^{cp}} + \underbrace{c_3\dot{p}}_{\dot{\boldsymbol{\epsilon}}^{up}}$$

With these preliminary remarks we can write down Leblond's TRIP model.

$$(2.3.34) \qquad \sigma_u = (1 - f_{cp}(p))\sigma_\gamma^y + f_{cp}(p)\sigma_\alpha^y$$

$$(2.3.35) \qquad \dot{\boldsymbol{\epsilon}}^{cp} = \frac{3(1-p)g(p)}{2\sigma_\gamma^y E}\boldsymbol{\sigma}^*\dot{\sigma}_{vm} + \frac{3(\lambda_\gamma - \lambda_\alpha)}{\sigma_\gamma^y}p\ln(p)\boldsymbol{\sigma}^*\dot{\theta} \qquad \text{if } \sigma_{vm} \leq \sigma_u$$

$$(2.3.36) \qquad \dot{\boldsymbol{\epsilon}}^{up} = -\frac{3\Delta\varepsilon_{\gamma\to\alpha}}{\sigma_\gamma^y}h\left(\frac{\sigma_{vm}}{\sigma_u}\right)\boldsymbol{\sigma}^*\ln(p)\dot{p} \qquad \text{if } \sigma_{vm} \leq \sigma_u$$

$$(2.3.37) \qquad \dot{\boldsymbol{\epsilon}}^p = \lambda\boldsymbol{\sigma}^* \qquad \text{if } \sigma_{vm} = \sigma_u$$

where σ_u is the so-called ultimate stress, $\sigma_\gamma^y, \sigma_\alpha^y$ is the yield stress of the γ (existent) or α (forming) phase. f_{cp} is a convex function, which was determined in experiments, see [LMD86b], just as function g. σ_{vm} is the equivalent strain by von Mises. $\lambda_\gamma, \lambda_\alpha$ are the thermal expansion coefficients of the respective phases, $\Delta\varepsilon_{\gamma\to\alpha}$ is the relative linear expansion caused by the phase change. h is a correction function for big stresses.

In [LMD89b] this model was extended to isotropic and kinematic hardening of austenite. The advantage of Leblond's model is that it needs only few material parameters which makes it easy to use. But there are also several disadvantages, as for example incapability of choosing the Greenwood-Johnson parameter stress dependent or the interaction of transformation induced plasticity and classical plasticity. For details see also [WBH08, WBS09].

2.4 Coupled problem

In this section we want to present the fully coupled problem. So mainly we gather the above presented equations but we also introduce some new coupling terms which could not be dealt with in the single effects subsections. We present the equations with emphasis on the dependencies between the different quantities.

At first we look at the heat equation, which is coupled with the phase transformation. When a material changes its state there is a certain amount of heat released or consumed, the so-called latent heat, L. During the quenching of steel austenite transforms to pearlite or martensite and so we have to consider an inner heat source on the right hand side of the heat equation. In the heat equation ρ_0 is the density of the material at the initial state. We do not consider dissipation, the transformation of mechanical energy into heat energy, otherwise there would be additional terms on the right hand side of the heat equation, compare e.g. [Wol08] for a detailed deduction of dissipation terms.

2.4.1. Heat equation:

$$\rho_0 c_e(\theta, P)\frac{\partial \theta}{\partial t} - \operatorname{div}(\kappa(\theta, P)\nabla\theta) = \rho_0 \sum_{i=2}^{m} L_i \dot{p}_i \qquad in\ \Omega \times (0,T)$$

$$\theta(\boldsymbol{x},0) \equiv \theta_0(\boldsymbol{x}) \qquad in\ \Omega$$

$$-\kappa(\theta, P)\nabla\theta \cdot n = \delta(\boldsymbol{x})(\theta - \theta_{ext}(\boldsymbol{x})) \qquad on\ \partial\Omega \times (0,T)$$

In the next step we present the phase transition equations for diffusion controlled and for martensitic transformations, compare equations (2.2.1, 2.2.2). Both models are understood to be non-reversible, the given equations apply only within the temperature region where the respective phases are forming. For simplicity we specify the models for the case of one forming phase, as we are exclusively considering such applications later on. The multi-phase case can be found in [WBB07]. Other phases which already formed lead to a reduced maximal possible phase fraction $\bar{p}$ in the Jonson-Mehl-Avrami equations and in the Koistinen-Marburger equation the expression is multiplied by the maximal possible phase fraction at the time, when the transformation starts.

2.4.2. Phase transition:

for one forming phase,
either the Johnson-Mehl-Avrami model for pearlite or bainite,

$$\dot{p}_2(\theta) = -(p_2(\theta) - \bar{p}(\theta))\frac{n(\theta)}{\tau(\theta)}\left(-\ln\left(1 - \frac{p_2(\theta)}{\bar{p}(\theta)}\right)\right)^{1-\frac{1}{n(\theta)}}$$

or the Koistinen-Marburger model for martensite,

$$p_3(\theta) = 1 - \exp^{-\frac{\theta_{ms}-\theta}{\theta_{m0}}}$$

see [WBB07] for the multi-phase case.

Finally we consider the deformation behaviour which is thermoelastoplasticity with transformation induced plasticity. The deformation equation itself remains unchanged while the strain and the stress tensor get new terms. The strain tensor is now split into three parts: an elastic, a plastic and a transformation induced plasticity part. This affects directly the form of the stress tensor which also gets a new term considering the stress caused by density changes due to quenching and phase transition. In this new term the so-called bulk modulus $K = \lambda + \frac{2}{3}\mu$ is introduced. In the flow function the elastic range $R_0 + R$ is now dependent on the temperature and the phases, which can be seen in the equation for the hardening. The back-stress model of Prager gains an additional term as the parameter c_{cp} becomes temperature dependent. The parameter c_{cp} depends on the parameters used for the modelling of isotropic hardening in the Ramberg-Osgood model, c_i, m_i.

Remark. Contrary to other approaches, here $\boldsymbol{\epsilon}^{el}$ incorporates also isotropic volume changes due to temperature changes and phase transitions.

2.4.3. Thermoelastoplasticity with TRIP:

- *Deformation equations*

$$\rho_0 \frac{\partial^2 \boldsymbol{u}}{\partial t^2} - \operatorname{div}(\boldsymbol{\sigma}) = \rho_0 \boldsymbol{f} \qquad\qquad in\ \Omega \times [0, T]$$

$$\boldsymbol{u}(0) = 0\ and\ \frac{\partial \boldsymbol{u}}{\partial t}(0) = 0 \qquad\qquad in\ \Omega$$

$$\boldsymbol{u} = g_D \qquad\qquad on\ \Gamma_D \times [0, T]$$

$$\boldsymbol{\sigma} \cdot \boldsymbol{n} = g_N \qquad\qquad on\ \Gamma_N \times [0, T]$$

- *Composition of the strain tensor and new form of the stress tensor*

$$\boldsymbol{\epsilon}(\boldsymbol{u}) = \boldsymbol{\epsilon}^{el} + \boldsymbol{\epsilon}^{cp} + \boldsymbol{\epsilon}^{up}$$

$$\boldsymbol{\sigma}(\boldsymbol{u}) = 2\mu(\theta, P)(\boldsymbol{\epsilon}(\boldsymbol{u}) - \boldsymbol{\epsilon}^{cp} - \boldsymbol{\epsilon}^{up}) +$$

$$+ \left(\lambda(\theta, P) \operatorname{tr}(\boldsymbol{\epsilon}(\boldsymbol{u})) - \left(K(\theta, P) \frac{\rho_0 - \rho(\theta, P)}{\rho(\theta, P)} \right) \right) \boldsymbol{I}$$

- *Yield criterion*

$$F = ||\boldsymbol{\sigma}^*(\boldsymbol{u}) - \boldsymbol{X}^{cp}|| - \sqrt{\frac{2}{3}} \Big(R_0(\theta, P) + R(\theta, P) \Big) = 0$$

- *Plastic flow rule*

$$\dot{\boldsymbol{\epsilon}}^{cp} = \gamma \frac{\boldsymbol{\sigma}^*(\boldsymbol{u}) - \boldsymbol{X}^{cp}}{||\boldsymbol{\sigma}^*(\boldsymbol{u}) - \boldsymbol{X}^{cp}||}$$

- *Hardening laws: either Ramberg-Osgood and Prager model or Armstrong-Frederick model*

  - *Ramberg-Osgood and Prager model*

$$\dot{s}^{cp} = \sqrt{\frac{2}{3}}\gamma \quad R = \sum_{i=1}^{m} p_i R_i \quad R_i = c_i(\theta)(s^{cp})^{m_i(\theta)}$$

$$\dot{\boldsymbol{X}}^{cp} = \frac{2}{3} c_{cp}(\theta) \dot{\boldsymbol{\epsilon}}^{cp} + \frac{2}{3} \frac{d}{dt} c_{cp}(\theta) \boldsymbol{\epsilon}^{cp} + \frac{2}{3} c_{int} \dot{\boldsymbol{\epsilon}}^{up}$$

$$\dot{\boldsymbol{X}}^{up} = \frac{2}{3} c_{int} \dot{\boldsymbol{\epsilon}}^{cp} + \frac{2}{3} c_{up} \dot{\boldsymbol{\epsilon}}^{up}$$

  (where $c_{int}, c_{up} = const$ and for a fixed s_{cp}:
  $c_{cp}(\theta) = \frac{d}{dt} R(\theta, s_{cp}))$

$-$ *Armstrong-Frederick model*

$$\dot{s}^{cp} = \sqrt{\frac{2}{3}}\gamma \quad R = \sum_{i=1}^{m} p_i R_i \quad \dot{R}_i = \hat{\gamma}_i \dot{s}^{cp} - \hat{\beta}_i R_i \dot{s}^{cp}$$

$$\dot{\boldsymbol{X}}^{cp} = \frac{2}{3}\hat{c}\dot{\boldsymbol{\epsilon}}^{cp} - \hat{b}\boldsymbol{X}^{cp}\dot{s}^{cp} + \frac{2}{3}c_{int}\dot{\boldsymbol{\epsilon}}^{up}$$

$$\dot{\boldsymbol{X}}^{up} = \frac{2}{3}c_{int}\dot{\boldsymbol{\epsilon}}^{cp} - c_{int}\frac{\hat{b}}{\hat{c}}\boldsymbol{X}^{cp}\dot{s}^{cp} + \frac{2}{3}c_{up}\dot{\boldsymbol{\epsilon}}^{up}$$

(where $\hat{\beta}_i = \hat{\beta}_i(\theta), \hat{b} = \hat{b}(\theta), \hat{c}, \hat{\gamma}_i, c_{int}, c_{up} = const$)

- *Evolution of TRIP*

$$\dot{\boldsymbol{\epsilon}}^{up} = (\boldsymbol{\sigma}^*(\boldsymbol{u}) - \boldsymbol{X}^{up}) \sum_{i=2}^{m} \frac{3}{2}\kappa^{up}(p_i)\dot{\Phi}(p_i)\dot{p}_i$$

Remark. It is also possible to formulate the new term in the stress tensor, which represents the density changes caused by quenching and phase transition, using the thermal expansion coefficient, α:

$$\left(K(\theta, P)\frac{\rho_0 - \rho(\theta, P)}{\rho(\theta, P)}\right)\boldsymbol{I} = \left(3\alpha(P)(\theta - \theta_0) - K(\theta, P)\sum_{i=1}^{m}\left(\frac{\rho_0}{\rho_i(\theta_0)} - 1\right)\right)\boldsymbol{I}$$

Remark. In the thermodynamical part of the modelling, which is not explained here, a function for the inelastic free energy is found, using inner variables describing isotropic and kinematic hardening. The hardening laws are then obtained by derivation of the inelastic free energy function. We demand, that the inelastic free energy is convex for frozen temperature and phases, hence the constants in the kinematic hardening models have to satisfy the following conditions:

(2.4.1)   Ramberg-Osgood and Prager model: $c_{int}^2 \le c_{up}c_{cp}$

(2.4.2)   Armstrong-Frederick model: $c_{int}^2 \le c_{up}\hat{c}$

Remark. In [WBMS] a more general coupling between the back-stresses of classical plasticity and transformation induced plasticity is described. This type of coupling has an additional term:

$$\dot{\boldsymbol{X}}^{cp} = \frac{2}{3}c_{cp}\dot{\boldsymbol{\epsilon}}^{cp} - a_{cp}\boldsymbol{X}^{cp}\dot{s}^{cp} + \frac{2}{3}c_{int}\dot{\boldsymbol{\epsilon}}^{up} - c_{int}\frac{a_{up}}{c_{up}}\boldsymbol{X}^{up}\dot{s}^{up}$$

$$\dot{\boldsymbol{X}}^{up} = \frac{2}{3}c_{int}\dot{\boldsymbol{\epsilon}}^{cp} - c_{int}\frac{a_{cp}}{c_{cp}}\boldsymbol{X}^{cp}\dot{s}^{cp} + \frac{2}{3}c_{up}\dot{\boldsymbol{\epsilon}}^{up} - a_{up}\boldsymbol{X}^{up}\dot{s}^{up}$$

$$\dot{s}^{up} = \sqrt{\frac{3}{2}}||\boldsymbol{\sigma}^* - \boldsymbol{X}^{up}||\sum_{i=2}^{m}\kappa^{up}(p_i)\dot{\Phi}(p_i)\dot{p}_i$$

We do not consider this model because there is no experimental data available for the determination of parameter a_{up} and because the numerical algorithm for the solution of the problem would become considerably more complicated, as is pointed out in [WBMS].

2.5   Weak formulation

After the formulation of the coupled system in the Section 2.4, the next step both for numerical as well as for analytical discussion is to transform the problem into its weak form. To obtain the weak formulation for the heat equation and the deformation equation as they are described in the Boxes 2.4.1 and 2.4.3 is a standard procedure. First the equation is multiplied by a test function, then it is integrated over the domain Ω followed by partial integration and some basic manipulations. Information on the function spaces used in this section can be found in the standard literature on functional analysis or in [Sho96].

In the weak formulation there occur several material parameters (e.g. c_e, κ), which depend on the solution itself. In the following Section 2.6, we will discuss shortly the problems of the lack of knowledge of existence and uniqueness of solutions of our coupled problem. For now we assume that the material parameters evaluated at θ, P are in $L^\infty([0,T] \times \Omega)$.

2.5.1. Weak heat equation:

We search a function
$$\theta \in V(0,T) := \left\{ f | f \in L^2(0,T; H^{1,2}(\Omega)), \frac{df}{dt} \in L^2(0,T; H^{1,2}(\Omega)^*) \right\}$$
for which holds
$$\theta(0) = \theta_0, \quad \theta_0 \in L^2(\Omega)$$

---

and for almost all $t \in (0, T)$ and all $\varphi \in H^{1,2}(\Omega)$:

$$(2.5.1) \quad \left\langle \rho_0 c_e \frac{d\theta}{dt} \middle| \varphi \right\rangle + \int_\Omega (\kappa \nabla \theta) \cdot \nabla \varphi dx + \int_{\partial\Omega} \delta\theta\varphi ds =$$

$$\int_\Omega \rho_0 \sum_{i=1}^m L_i \dot{p}_i \varphi dx + \int_{\partial\Omega} \delta\theta_{ext}\varphi ds$$

where:

$\rho_0 > 0, \quad p_i \in L^\infty([0,1]), \quad \dot{p}_i \in L^2(0,T;L^2(\Omega)),$
$c_e(\theta,P), \kappa(\theta,P) \in L^\infty([0,T] \times \Omega), \quad \nabla c_e(\theta,P) \in L^\infty([0,T] \times \Omega)^3,$
$\delta \in L^\infty(0,T;L^\infty(\partial\Omega)), \quad \theta_{ext}(t,\boldsymbol{x}) \in L^2(0,T;L^2(\partial\Omega)).$

In the weak formulation of the deformation equation we do not consider the body force $\boldsymbol{f}$ as it does not occur in our problem setting. It is also worth noting that in most cases it is sensible to neglect the inertia term $\frac{\partial^2 \boldsymbol{u}}{\partial t^2}$ because we are usually not discussing situations were stresses appear or vanish abruptly, compare Section 4.3. In neglecting the inertia term the partial differential equation changes its type from hyperbolic to elliptic. Henceforth we consider a pure Neumann-problem and change the name of the function on the right hand side from $\boldsymbol{g}_n$ to $\boldsymbol{F}_{ext}$.

2.5.2. Weak deformation equation:

We search a function

$$\boldsymbol{u} \in W(0,T) := \left\{ f \middle| f \in L^2(0,T;(H^{1,2}(\Omega))^3), \frac{df}{dt} \in L^2(0,T;L^2(\Omega)), \right.$$

$$\left. \frac{d^2 f}{dt^2} \in L^2(0,T;(H^{1,2}(\Omega)3)^*) \right\}$$

for which holds

$$\boldsymbol{u}(0) = 0 \ a.e., \ and \ \frac{\partial\boldsymbol{u}}{\partial t} = 0 \ a.e.$$

and for almost all $t \in (0,T)$ and all $\boldsymbol{\phi} \in (H^{1,2}(\Omega))^3$:

$$(2.5.2) \quad \left\langle \rho_0 \frac{\partial^2 \boldsymbol{u}}{\partial t^2} \middle| \boldsymbol{\phi} \right\rangle + \int_\Omega 2\mu\boldsymbol{\epsilon}(\boldsymbol{u}) : \boldsymbol{\epsilon}(\boldsymbol{\phi}) + \lambda \operatorname{div}(\boldsymbol{u})\operatorname{div}(\boldsymbol{\phi})dx =$$

$$\int_\Omega K\left(\frac{\rho_0 - \rho}{\rho}\right)\boldsymbol{I} : \nabla\boldsymbol{\phi}dx + \int_\Omega 2\mu(\boldsymbol{\epsilon}^{up} + \boldsymbol{\epsilon}^{cp}) : \nabla\boldsymbol{\phi}dx + \int_{\partial\Omega} \boldsymbol{F}_{ext} \cdot \boldsymbol{\phi}ds$$

where:

$\mu(\theta,P), \lambda(\theta,P), K(\theta,P) \in L^{\infty}([0,T] \times \Omega),$

$\rho(\theta,P) \in L^{\infty}([0,T] \times \Omega), \quad \rho(\theta,P) \geq c > 0 \; a.e.,$

$\boldsymbol{F}_{ext} \in L^2(0,T;L^2(\partial\Omega))^3,$

$\boldsymbol{\epsilon}^{up}, \boldsymbol{\epsilon}^{cp} \in L^2(0,T;L^2(\Omega))^5$

The formulation of the partial differential equation in their weak form is essential for the numerical treatment of the problem in the next chapter, for the ordinary differential equations this is not necessary. With these prerequisites we created the basis for some remarks on the problems of existence and uniqueness of solutions to the problems in the next section.

2.6   Remarks on existence and uniqueness of solutions

As the title of this section already indicates, there are neither proofs on the existence nor on the uniqueness of solutions to the full coupled problem introduced in the preceding sections. Nevertheless there exist results for similar or (sub)problems, where usually some coupling or parameter dependencies are dropped. In this section we will name only some results on this field which are very closely related to our considered problem, as the main aspect of this thesis lies in the numerical treatment of the problem.

To start with, the group of Hömberg published several papers on the modelling and optimisation of phase transitions in steel. In [Höm96, Höm97] they deal with a setting where, in addition to heat conduction, there occur phase transformations from austenite to pearlite and from austenite to martensite. Those phase transitions are modeled differently from our approach and include Heaviside functions (e.g. to take into account the irreversibility of the process). It is proved that under certain conditions (e.g. constant material parameters) there exist a unique solution, $\theta_{\delta} \in H^{2,1}(\Omega) = H^1(0,T;L^2(\Omega)) \cap L^2(0,T;H^1(\Omega))$, for a regularised heat equation without Heaviside functions. Then the existence of a solution triple, $(\theta, v, w) \in H^{2,1}(\Omega) \times L^{\infty}(\Omega) \times L^{\infty}(\Omega)$, for the general case is proved, where v and w replace the Heaviside functions in the phase transformation models.

In [Höm04] Hömberg deals with a model for induction hardening of steel including mechanical effects and shows under which restrictions there exits a weak solution to the coupled system. Besides other effects, it deals with heat conduction, phase transitions, transformation strain and transformation induced plasticity. Relating to previous works on heat conduction and phase transition, the author assumes the relation between temperature evolution and phase fraction to be known a priori. Other restrictions for the existence of a weak solution affect the material parameters. Many of them are assumed to be continuous or continuously differentiable and their physical dependence on temperature, phase fraction, time etc. is limited. The quantities, which are important for us, are $\theta \in L^q(0,T;W^{1,q}(\Omega))$, $\sigma \in L^2(0,T;\{(\sigma_{ij})|\sigma_{ij} = \sigma_{ji}$, and $\sigma_{ij} \in L^2(\Omega)\, i,j = 1,2,3\})$ and $u \in L^2(0,T;(H^1(\Omega))^3)$.

In her Diploma-thesis [Hüß07], Hüßler deals with a coupled problem of heat conduction, phase transformation and carbon diffusion. It is proved under which conditions there exist (unique) weak solutions in Banach- or Hilbert-spaces for the single aspects and the coupled problem. In a special case, where only Bainite and Martensite are forming, it is proved under which conditions the existence of a solution triple of temperature, carbon fraction in the austenitic phase and phase fractions, $(\theta, u_{c1}, p) \in H^{1,2}(0,T;V,H) \times H^{1,2}(0,T;V,H) \times \{f \in L^\infty(\Omega \times (0,T))|\dot{f} \in L^\infty(\Omega \times (0,T))\}$, holds.

In another Diploma-thesis [Böt07] Böttcher deals with thermoelasticity with phase transitions and transformation induced plasticity. After dealing with existence and uniqueness of weak solutions for stationary and nonstationary linear elasticity, linear thermoelasticity with different boundary conditions is treated. Finally existence and uniqueness of solutions of the coupled problem thermoelasticity, phase transitions and transformation induced plasticity is proved.

For linear elasticity there are plenty of results on existence and uniqueness of strong or weak solutions, we refer to [Böt07, Bra07, Gei04].

Considering plasticity some results have been achieved by Johnson. In [Joh76] he deals with elastic perfectly plastic material, using a Prandtl-Reuss flow rule and considering time-dependent loads. A mixed formulation (compare Section 3.1) including stress and displacement is used and the problem is formulated as variational inequality which is proved to have a strong solution $(\sigma, v), \sigma \in L^\infty(0,T;[L^2(\Omega)]^6), \dot{\sigma} \in L^2(0,T;[L^2(\Omega)]^6), v \in L^2(0,T;[L^{2/3}(\Omega)]^3)$. Analogous work for plasticity with isotropic and kinematic hardening is carried out in [Joh78].

In their book [HR99], Han and Reddy deal with classical plasticity with isotropic and linear kinematic hardening in the variational formulation. They prove existence and uniqueness of solutions for both primal and dual variational problem.

Chełmiński and Racke deal in [CR06] with a problem of thermoplasticity with a modified Prandtl-Reuss flow rule and linear kinematic hardening. They replaced the flow rule by its Yosida approximation and modified the right hand side of the heat equation. After proving the existence of solutions of the approximated system, they finally proved the existence of global in time strong L^2-solutions of thermoplasticity with linear kinematic hardening.

CHAPTER

3

Discretisation of the problem

In the last chapter we formulated the coupled system of equations which describes the heat treatment problem together with its weak formulation. Here we will develop and describe in detail a scheme for the numerical solution of the heat treatment problem. First we will point out briefly that there are different approaches for dealing numerically with the deformation equation. After giving an overall solution scheme for the complete problem in the second section, we will then describe the solution of different types of equations in detail in the three following sections. For the discretisation of the two PDEs, the heat and the deformation equation, we will use the Finite Element method. As the Finite Element method (FEM) is well-known and widespread we will not explain it here but refer to the standard literature or, as we use their notations, to [Bra07, SS05]. In the sixth section we will explain how special boundary conditions like periodicity or symmetry are implemented into the program.

In the last section of this chapter we will describe in detail the a posteriori error indicators which are used for the adaptive Finite Element computations.

3.1   Approaches for the deformation equation

In the literature there are multiple approaches to deal with the deformation equation numerically. For demonstration we will consider the problem of linear elasticity, as described by equations (2.3.5) and (2.3.6), but without taking the inertia term into account. Even in this relatively simple case, there are several approaches which differ in the question whether displacement, strain and stress are treated as own variables or are expressed via one or two of the others (using the respective equations). Three different approaches are relatively widespread and will be explained in this section in brevity, compare [Bra07, Gei04] for details.

The displacement approach Here both stress and strain are expressed through the displacement. This simplifies the solution of the deformation equation itself:

$$-\operatorname{div}\left(\mu(\nabla \boldsymbol{u} + \nabla \boldsymbol{u}^T) + \lambda \sum_{i=1}^{3} \frac{\partial u_i}{\partial x_i} \boldsymbol{I}\right) = \boldsymbol{f} \qquad \text{in } \Omega$$

$$\boldsymbol{u} = g_D \qquad \text{on } \Gamma_D$$

$$\left(\mu(\nabla \boldsymbol{u} + \nabla \boldsymbol{u}^T) + \lambda \sum_{i=1}^{3} \frac{\partial u_i}{\partial x_i} \boldsymbol{I}\right) \cdot \boldsymbol{n} = g_N \qquad \text{on } \Gamma_N$$

On the other hand one is often interested in the stress, e.g. we need it for the computation of TRIP and classical plasticity. The stress can only be obtained by post-processing, causing more computational effort:

$$\boldsymbol{\sigma} = \mu(\nabla \boldsymbol{u} + \nabla \boldsymbol{u}^T) + \lambda \sum_{i=1}^{3} \frac{\partial u_i}{\partial x_i} \boldsymbol{I}$$

The mixed method of Hellinger and Reissner In this method the displacement and the stress are kept as variables while the strain is expressed through the displacement.

$$-\operatorname{div}(\boldsymbol{\sigma}) = \boldsymbol{f} \qquad \text{in } \Omega$$

$$\boldsymbol{\sigma} = \mu(\nabla \boldsymbol{u} + \nabla \boldsymbol{u}^T) + \lambda \sum_{i=1}^{3} \frac{\partial u_i}{\partial x_i} \boldsymbol{I} \qquad \text{in } \Omega$$

$$\boldsymbol{u} = g_D \qquad \text{on } \Gamma_D$$

$$\boldsymbol{\sigma} \cdot \boldsymbol{n} = g_N \qquad \text{on } \Gamma_N$$

The mixed method of Hu and Washizu   Here all quantities, displacement, stress and strain stay as variables within the equation.

$$
\begin{aligned}
- \operatorname{div}(\boldsymbol{\sigma}) &= \boldsymbol{f} && \text{in } \Omega \\
\boldsymbol{\sigma} &= 2\mu\boldsymbol{\epsilon} + \lambda \operatorname{tr}(\boldsymbol{\epsilon})\boldsymbol{I} && \text{in } \Omega \\
\boldsymbol{\epsilon} &= \frac{1}{2}(\nabla \boldsymbol{u} + \nabla \boldsymbol{u}^{T}) && \text{in } \Omega \\
\boldsymbol{u} &= g_D && \text{on } \Gamma_D \\
\boldsymbol{\sigma} \cdot \boldsymbol{n} &= g_N && \text{on } \Gamma_N
\end{aligned}
$$

It is not a priori clear which method is best for which problem as each of the three has advantages and disadvantages in respect to convenience of the implementation, properties of the matrix (of the resulting system of linear equations when FEM is applied) and error estimates of the adaptive FEM. A more detailed introduction to all three approaches can be found in [Gei04], while proofs and remarks concerning the existence and uniqueness of solutions can be found in [Bra07].

In the following we will use the displacement method, because the resulting system matrix has favourable properties (symmetrical positive definite when there is a Dirichlet boundary condition). Besides this, the displacement method is relatively easy to implement.

Remark. In the numerical solution of the problem of linear elasticity an effect called *locking* can occur. There are different kinds of locking like volume locking in nearly incompressible materials or shear locking in beams. The problem of volume locking is, that when the Poisson ratio $\nu \to 0.5$ the numerical solution approximated by low order elements does not converge uniformly towards the exact solution. An explicative derivation of locking in linear elasticity can be found in [Bra07]. Here also three possibilities of avoiding the locking effect are mentioned, namely formulating a saddle point problem with penalty term, reduced integration and assumed strain methods. Another method of overcoming locking is the use of mixed formulations. A literature review on works on locking can be found in [DLRW06]. This Paper deals with conditions for equivalence of several approaches as the low order displacement approach, the mixed Hu-Washizu formulation, the method of enhanced strains and the method of enhanced assumed stresses. Numerical simulations comparing these different approaches (with and without locking) are presented, too.

3.2   Overall numerical solution scheme

After choosing an approach for the deformation equation, we will now start to develop a scheme which allows us to solve numerically our coupled problem introduced in Section 2.4. As our setting contains several partial and ordinary differential equations which are multiply coupled, we have to make some simplifications. To consider the full coupling, as presented in Section 2.4, would result in a completely implicit algorithm and would raise the computational effort beyond any reason.

Based on a FEM-code of Siebert [MNS] for elasticity, Schmidt developed a code for the the coupled problem of the heat equation, phase transition, elasticity and a simple approach of transformation induced plasticity [SWB03]. For this thesis this code was extended with different models for phase transformation, a more general model for transformation induced plasticity and classical plasticity, including different hardening models and an additional term in the time step size control. Also the periodic and symmetric boundary conditions were added.

The already mentioned overall numerical scheme for the solution of our coupled problem will describe briefly in which order and with which coupling the equations will be solved. In some cases a so-called weak coupling is applied, meaning that values from the last time step will be used. This procedure facilitates the solution of the problem considerably, while one can assume that the error will be small for sufficiently small time steps.

First of all we will discretise or solution interval $[0, T]$ into time steps $t_0 = 0, t_1, \ldots,$ $t_m = T$. The following scheme describes the course of action in one single time step.

3.2.1. Overall numerical solution scheme:

Start with initial values for t_0: $\theta_0, p_0, \epsilon_0^{cp}, \epsilon_0^{up}$
All quantities are known for t_{n-1}, compute their values for t_n:

1. *compute θ_n by solving the heat equation via FEM, using old values for phase fraction and its derivative, $p_{n-1}, \dot{p}_{n-1}$, see Section 3.3*

2. *obtain p_n: solve ODEs for phase transition via Euler-scheme, using updated temperature θ_n, see Section 3.4*

---

3. *compute $\boldsymbol{u}_n$ by solving the deformation equation via FEM, using updated temperature, θ_n, and phase fraction, p_n, but TRIP and plasticity tensors from the previous time step, $\boldsymbol{\epsilon}^{up}_{n-1}, \boldsymbol{\epsilon}^{cp}_{n-1}$, see Section 3.3*

4. *if necessary, update classic-plastic quantities, $F_n, \gamma_n, \boldsymbol{\epsilon}^{cp}_n, \boldsymbol{X}^{cp}_n, s^{cp}_n, R_n$, using the algorithm explained in Section 3.5 (otherwise, the quantities stay constant)*

5. *if necessary, solve TRIP ODE via Crank-Nicolson scheme, gain $\boldsymbol{\epsilon}^{up}_n, \boldsymbol{X}^{up}_n$, see Section 3.4 (otherwise, the quantities stay constant)*

The details about the solution procedure will be explained in the following three sections. But first, we have to formulate some necessary prerequisites.

As we will apply the Finite Element Method to our problem we will now specify which quantities are approximated in which Finite Element spaces.

For the spatial discretisation of Ω we choose a shape regular, simplicial triangulation $\mathcal{T}_h$. So in three dimensions our triangulation consists of tetrahedrons which have a maximal edge length h. Then we use piecewise polynomial Lagrange Finite Elements φ for the definition of the corresponding FE-space on $\mathcal{T}_h$.

All scalar quantities, like $\theta_n, p_n, F_n, \gamma_n, s^{cp}_n, R_n$, are approximated as linear combinations of our scalar basis:

$$(3.2.1) \qquad V_h = \operatorname{span}\{\varphi_1, \ldots, \varphi_N\} \subset V := H^{1,2}(\Omega)$$

which is of dimension N (The dimension of V_h equals its number of degrees of freedom).

As the deformation $\boldsymbol{u}_n$ is a three-dimensional vector field, we define its FE-space,

$$(3.2.2) \qquad V^3_h = \operatorname{span}\left\{ \begin{pmatrix} \varphi_i \\ 0 \\ 0 \end{pmatrix}, \begin{pmatrix} 0 \\ \varphi_i \\ 0 \end{pmatrix}, \begin{pmatrix} 0 \\ 0 \\ \varphi_i \end{pmatrix} \right\}_{i=1,\ldots,N}$$
$$= \operatorname{span}\{\boldsymbol{\phi}_1, \ldots, \boldsymbol{\phi}_{3N}\} \subset W := (H^{1,2}(\Omega))^3$$

which is of dimension $3N$.

In our setting all tensors, $\boldsymbol{\sigma}, \boldsymbol{\epsilon}^{up}_n, \boldsymbol{\epsilon}^{cp}_n, \boldsymbol{X}^{cp}_n, \boldsymbol{X}^{up}_n$, are symmetric. Hence it will save considerably memory when we define the FE-space for symmetric tensors as follows:

$$(3.2.3) \qquad V^6_h = \operatorname{span}\left\{ \boldsymbol{\Phi}^i_{kl}, k,l = 1,2,3, k \leq l \right\}_{i=1,\ldots,N},$$

where diagonal and non-diagonal elements are defined as follows:

$$(3.2.4) \qquad \boldsymbol{\Phi}^i_{11} = \begin{bmatrix} \varphi_i & 0 & 0 \\ 0 & 0 & 0 \\ 0 & 0 & 0 \end{bmatrix}, \qquad \boldsymbol{\Phi}^i_{12} = \frac{1}{\sqrt{2}} \begin{bmatrix} 0 & \varphi_i & 0 \\ \varphi_i & 0 & 0 \\ 0 & 0 & 0 \end{bmatrix}, \text{etc}$$

Our FE-space V^6_h is of dimension $6N$.

Remark. The factor $\frac{1}{\sqrt{2}}$ for the non-diagonal entries was chosen, such that:
$\left(\boldsymbol{\Phi}^i_{kl}\middle|\boldsymbol{\Phi}^j_{kl}\right)_{L^2(\Omega)} = \int_\Omega \varphi_i \varphi_j dx$ is satisfied.

Remark. Many tensors present in our problem are traceless. The condition $\boldsymbol{\Phi}^i_{33} = -\boldsymbol{\Phi}^i_{11} - \boldsymbol{\Phi}^i_{22}$ allows us again to save memory by using the smaller FE-space

$$(3.2.5) \qquad V^5_h = \text{span}\left\{\boldsymbol{\Phi}^i_{kl}, k = 1, 2; l = 1, 2, 3; k \le l\right\}_{i=1,\dots,N}$$

of dimension $5N$ for the traceless tensors.

By now we collected all necessary prerequisites and can go on with the detailed description of the single steps of the overall solution scheme.

3.3 Application of the Finite Element method

As already mentioned before, we are not going to explain the Finite Element method itself, but only describe its application to the heat equation and the deformation equation.

As a first step we look at the time discretisation of the heat equation. For this we employ a ϑ-method, meaning that the equation

$$\frac{d\theta}{dt} = F\left(\boldsymbol{x}, t, \theta, \frac{\partial\theta}{\partial\boldsymbol{x}}, \frac{\partial^2\theta}{\partial\boldsymbol{x}^2}\right)$$

is discretised by

$$\frac{\theta_n - \theta_{n-1}}{\tau_n} = \vartheta F_n\left(\boldsymbol{x}, t, \theta, \frac{\partial\theta}{\partial\boldsymbol{x}}, \frac{\partial^2\theta}{\partial\boldsymbol{x}^2}\right) + (1-\vartheta)F_{n-1}\left(\boldsymbol{x}, t, \theta, \frac{\partial\theta}{\partial\boldsymbol{x}}, \frac{\partial^2\theta}{\partial\boldsymbol{x}^2}\right)$$

By choosing $\vartheta = 0$ we have a explicit Euler scheme, by choosing $\vartheta = 1$ we have a implicit Euler scheme and by choosing $\vartheta = \frac{1}{2}$ we have a Crank-Nicolson scheme.

We want to compute an approximate solution $\theta_n \in V_h$ to the exact solution $\theta(t_n) \in V$.

Thus we set

$$(3.3.1) \qquad\qquad \theta_n = \sum_{i=1}^{N} \theta_n^i \varphi_i$$

and the aim is to calculate the coefficient vector $(\overrightarrow{\theta_n})_i = \theta_n^i$.

To obtain the time discretised heat equation we apply the above explained Crank-Nicolson scheme to the weak heat equation from Box 2.5.1 and choose $\varphi_j, j = 1, \ldots, N$ as test-functions.

$$(3.3.2) \quad \int_\Omega \rho_0 c_{e,n-1} \frac{\theta_n - \theta_{n-1}}{\tau_n} \varphi_j + \vartheta \kappa_{n-1} \nabla \theta_n \cdot \nabla \varphi_j dx + \int_{\partial\Omega} \delta \theta_n \varphi_j ds =$$

$$\int_\Omega (1 - \vartheta) \kappa_{n-1} \nabla \theta_{n-1} \cdot \nabla \varphi_j dx + \int_\Omega \rho_0 \sum_{k=2}^{m} L_{n-1}^k \dot{p}_{n-1}^k \varphi_j dx + \int_{\partial\Omega} \delta \theta_{n-1}^{ext} \varphi_j ds$$

$$\text{for } j = 1, \ldots, N$$

For the spatial discretisation we apply the Galerkin method to the time discretised heat equation. With equations (3.3.1) and (3.3.2) we obtain the fully discretised heat equation.

3.3.1. Discretised heat equation:

$$\sum_{i=1}^{N} \left(\frac{\rho_0}{\tau_n} \int_\Omega c_e \varphi_i \varphi_j dx - \vartheta \int_\Omega \kappa \nabla \varphi_i \nabla \varphi_j dx + \int_{\partial\Omega} \delta \varphi_i \varphi_j ds \right) (\overrightarrow{\theta_n})_i =$$

$$\sum_{i=1}^{N} \left(\frac{\rho_0}{\tau_n} \int_\Omega c_e \varphi_i \varphi_j dx + (1 - \vartheta) \int_\Omega \kappa \nabla \varphi_i \nabla \varphi_j dx \right) (\overrightarrow{\theta_{n-1}})_i +$$

$$+ \int_\Omega \rho_0 \sum_{k=2}^{m} L_{n-1}^k \dot{p}_{n-1}^k \varphi_j + \int_{\partial\Omega} \delta \theta_{n-1}^{ext} \varphi_j$$

$$\text{for } j = 1, \ldots, N$$

(Due to space restrictions, we dropped the notations $c_{e,n-1}$ and κ_{n-1}.)

In this way we achieve a system of linear equations of dimension $N \times N$, whose solution $(\overrightarrow{\theta_n})$ gives us θ_n: the approximation of $\theta(t_n)$ in V_h.

Remark. In this context one often uses the notion of the mass matrix $M_{n-1}^{ij} = \int_\Omega c_{e,n-1}\varphi_i\varphi_j dx$ and the stiffness matrix $A_{n-1}^{ij} = \int_\Omega \kappa_{n-1}\nabla\varphi_i\nabla\varphi_j dx$.

Remark. It is worth noting that if one chooses linear Finite Elements the usage of a so-called lumped scalar product for the mass matrix is helpful. It only considers values at the nodes of the triangulation and in this way diagonalises the mass matrix, compare [GR92] for details.

The above equation is solved in the first step of the overall solution scheme 3.2.1.

In the discretisation of the deformation equation we will at first disregard the inertia term. The problems we are considering are usually independent of the velocity of applied forces and in these cases the inertia term is irrelevant. In Section 4.3, we will discuss in detail cases where inertia forces have influence. By considering the so-called quasi-static case (3.3.3), our partial differential equation changes its type from hyperbolic to elliptic.

$$(3.3.3) \quad \int_\Omega 2\mu\epsilon(\boldsymbol{u}):\epsilon(\boldsymbol{\phi}) + \lambda\operatorname{div}(\boldsymbol{u})\operatorname{div}(\boldsymbol{\phi})dx = \int_{\partial\Omega} \boldsymbol{F}_{ext}\cdot\boldsymbol{\phi}ds+$$

$$\int_\Omega K\left(\frac{\rho_0-\rho}{\rho}\right)\boldsymbol{I}:\nabla\boldsymbol{\phi}dx + \int_\Omega 2\mu(\epsilon^{up}+\epsilon^{cp}):\nabla\boldsymbol{\phi}dx$$

As the above equation is stationary and the time acts only as a parameter, we can begin directly with the spatial discretisation which is done in the same way as for the heat equation. We set

$$(3.3.4) \qquad\qquad\qquad \boldsymbol{u}_n = \sum_{i=1}^{3N} \boldsymbol{u}_n^i\phi_i \in V_h^3$$

and want to find the coefficient vector $(\overrightarrow{\boldsymbol{u}_n})_i = \boldsymbol{u}_n^i$. When we plug this into (3.3.3), we obtain the discretised deformation equation.

3.3.2. Discretised (stationary) deformation equation :

$$\sum_{i=1}^{3N} \left(\int_{\Omega} 2\mu\epsilon(\boldsymbol{\phi}_i) : \epsilon(\boldsymbol{\phi}_j) + \lambda \operatorname{div}(\boldsymbol{\phi}_i) \operatorname{div}(\boldsymbol{\phi}_j) dx \right) \boldsymbol{u}_n^i = \int_{\partial\Omega} \boldsymbol{F}_{ext} \cdot \boldsymbol{\phi}_j ds +$$

$$\int_{\Omega} K \left(\frac{\rho_0 - \rho}{\rho} \right) \boldsymbol{I} : \nabla\boldsymbol{\phi}_j dx + \int_{\Omega} 2\mu(\epsilon^{up} + \epsilon^{cp}) : \nabla\boldsymbol{\phi}_j dx \quad, j = 1, \dots, 3N$$

In this way we achieve a linear system of equations of dimension $3N \times 3N$, whose solution $(\overrightarrow{\boldsymbol{u}_n})$ gives us $\boldsymbol{u}_n$: the approximation of $\boldsymbol{u}(t_n)$ in V_h^3.

This equation is solved in the third step of the overall solution scheme 3.2.1.

3.4 Calculation of the phase transformation and TRIP

In this section we will deal with the solution of the ordinary differential equations for the phase transformation and transformation induced plasticity.

The new phase fraction is calculated in step two of our overall solution scheme 3.2.1. The phase transformation models of Johnson-Mehl-Avrami (2.2.1) and Leblond (2.2.3) are formulated as ordinary differential equations. To increase the accuracy of the solution, the time step $\tau_n := t_n - t_{n-1}$ is subdivided into ten sub-time-steps. The temperature is interpolated linearly between t_{n-1} and t_n to gain its values at the sub-time-steps. Then the corresponding ordinary differential equations are solved, using an implicit Euler scheme.

Special cases are the Koistinen-Marburger equation and the model of Yu for the martensitic transformation. Here the phase fraction can be computed directly.

In the last step of our overall numerical scheme in Box 3.2.1 we have to check whether the phase fractions increased in this time step. If so, then we need to update the TRIP tensor, ϵ^{up}. The more complicated case of transformation induced plasticity with back-stress coupled to the classical plasticity is treated in the next section. Here we will deal with the simpler situation where classical plasticity is not taken into account. For the update of the TRIP tensor we need to calculate the stress tensor which involves the calculation of the strain tensor. As the strain tensor, $\epsilon_n = \frac{1}{2}(\nabla\boldsymbol{u}_n + \nabla\boldsymbol{u}_n^T)$, consists of the gradient of the deformation, its components are not element of the FE-space.

Therefore we project the strain tensor, component by component, onto the FE-space V_h:
$\tilde{\epsilon}_{kl} := P_{V_h}(\epsilon_{kl})$, $k, l = 1, 2, 3, k \leq l$. Because $\tilde{\epsilon}$ is the projection into V_h it holds:

(3.4.1a) $$\tilde{\epsilon}_{kl} \in V_h \qquad \tilde{\epsilon}_{kl}(x) = \sum_{i=1}^{N} (\tilde{\epsilon}_{kl})_i \varphi_i(x)$$

(3.4.1b) $$\int_{\Omega} \tilde{\epsilon}_{kl} \varphi dx = \int_{\Omega} \epsilon_{kl} \varphi dx$$

When we insert equation (3.4.1a) into equation (3.4.1b), we obtain a linear system of equations for the determination of the coefficients $(\tilde{\epsilon}_{kl})_i$.

(3.4.2) $$\sum_{i=1}^{N} \left(\int_{\Omega} \varphi_i \varphi_j dx \right) \overrightarrow{(\tilde{\epsilon}_{kl})}_i = \int_{\Omega} \epsilon_{kl} \varphi_j dx, \quad j = 1, \dots, N$$

This newly calculated projection of the strain tensor onto the FE-space, is used for the calculation of the stress tensor. To keep the notation simple, we drop the tilde from now on. Whenever we mention the strain tensor, its projection is meant.

(3.4.3) $$\boldsymbol{\sigma}_n = \left(\lambda(\theta_n, p_n) \text{tr}(\boldsymbol{\epsilon}(\boldsymbol{u}_n)) - K(\theta_n, p_n) \frac{\rho_0 - \rho(\theta_n, p_n)}{\rho(\theta_n, p_n)} \right) \boldsymbol{I} +$$
$$+ 2\mu(\theta_n, p_n) \left(\boldsymbol{\epsilon}(\boldsymbol{u}_n) - \boldsymbol{\epsilon}_{n-1}^{cp} - \boldsymbol{\epsilon}_{n-1}^{up} \right)$$

After these preliminaries we can update the TRIP tensor via a Crank-Nicholson scheme.

(3.4.4) $$\boldsymbol{\epsilon}_n^{up} = \boldsymbol{\epsilon}_{n-1}^{up} + \frac{1}{2}(\boldsymbol{\sigma}_n^* + \boldsymbol{\sigma}_{n-1}^*)\tau_n \frac{3}{2}\kappa \sum_{i=2}^{m} \dot{\Phi}_i(p_i)\dot{p}_i$$

In this way the last step of our overall numerical scheme in Box 3.2.1 is carried out in the case where classical plasticity is not considered.

3.5   Algorithms for plasticity

After step three in the overall solution scheme 3.2.1 it has to be checked whether this time step is elastic or plastic and, if necessary, the classical plasticity and transformation induced plasticity quantities have to be updated. In this section we will explain the case of plasticity with isotropic and kinematic hardening using the Ramberg-Osgood and Prager model. To avoid repetitions the equations for plasticity with isotropic and kinematic hardening using the Armstrong-Frederick model are given in Appendix A. The procedure of the derivation of the algorithm is similar but the resulting equations are more complicated.

The algorithm we use is based on an algorithm developed by Simo and Hughes [SH97]. In this work we will introduce a semi-implicit algorithm, which takes into account non-linear isotropic and kinematic hardening and also includes a coupling between the back stresses of classical plasticity and transformation induced plasticity. In the literature there can be found fully implicit algorithms for plasticity, but here simplifications were made in the hardening behaviour and a back stress coupling between classical plasticity can not be found.

Fully implicit algorithms consist usually of an outer and an inner iteration, compare for example [SH97]. In the outer iteration a new displacement $\boldsymbol{u}_{n+1}^k$ is calculated using the consistent tangent modulus, $\frac{\partial \boldsymbol{\sigma}}{\partial \boldsymbol{\epsilon}}$, introduced in [ST85]. The inner iteration updates all plastic quantities. More details on the fully implicit algorithm can be found in the box below.

3.5.1. Fully implicit iterative algorithm by Simo and Hughes:

Let $(\cdot)_{n+1}^k$ be the values of the variable $(\cdot)$ at time t_{n+1} at the k-th iteration.

- *Outer iteration:*
  determine a new $\boldsymbol{u}_{n+1}^k$ using the consistent tangent modulus

  - *Inner iteration:*
    start with a new displacement $\boldsymbol{u}_{n+1}^k$ and calculate strain tensor, (corrected) stress tensor and plasticity tensor via numerical solution of a system of equations

- *put these values into the deformation equation and calculate a residual r*

- **If** *r < tol then* $(\cdot)_{n+1}^k$ *is the solution* **DONE**

- **Else***:* **CONTINUE**

In [ST85] there was shown by Simo and Taylor, that the use of the consistent tangent modulus is preferable to the elasto-plastic tangent modulus, which often had been used before. Following their approach Hartmann and Haupt [HH93] introduced a fully implicit algorithm for plasticity with non-linear kinematic hardening using the principle of virtual work. Doghri dealt in [Dog93] with plasticity with non-linear isotropic and kinematic hardening. He specified the consistent tangent modulus for this problem and derived a scheme for the inner iteration, where a system of two equations for the plastic multiplier and the effective stress has to be solved. In [Mah99] Mahnken improved Doghri's scheme for the inner iteration and reduced the system of equations to one scalar equation for the plastic multiplier. Later in [MSA09] Mahnken et. al. dealt with the quenching of steel. Their setting is similar to ours: low-alloy steel transforms form austenite to martensite while classical plasticity and transformation induced plasticity are present. Contrary to our approach there is no kinematic hardening and thus no coupling between classical and transformation induced plasticity. For the inner iteration a scheme is specified, which involves only the solution of a scalar equation for the plastic multiplier. For the outer iteration the consistent tangent modulus for this setting is specified.

The method we use is not fully but semi-implicit, because we have no outer iteration. We use the old values of the plasticity and TRIP tensor to compute the new displacement. For the inner iteration we derive a scheme, so that for the correction of the stress tensor and the update of the plastic quantities only one scalar equation for the plastic multiplier has to be solved. Thereafter the displacement is not updated. A disadvantage of the semi-implicit method is that it needs smaller time steps than the implicit method. On the other hand for the implicit method the consistent tangent modulus is necessary, which would be very complicated to obtain for a classical plasticity with isotropic and kinematic hardening and interaction with transformation induced plasticity.

Remark. A different approach is to deal with classical plasticity as variational problem. In Han and Reddy [HR99], the solution of the elasto-plastic problem is treated as variational inequality. They discuss the analysis and numerics of the primal variational problem, as well as the dual variational problem. The primal variational problem is used e.g. by Carstensen and Alberty et. al. in [ACZ99, CA03, COV06] where, among other things, adaptivity and error estimates for elasto-plasticity are treated.

Now we will explain in detail our algorithm for dealing with classical plasticity with isotropic and kinematic hardening, including a coupling between classical plasticity and transformation induced plasticity.

All following quantities are space dependent. The calculations are conducted in each degree of freedom separately and can be understood as a post processing after the solution of the partial differential equation for the deformation. Therefore there is no coupling between neighbouring points.

In every time step, after the solution of the deformation equation, we have to check whether this step is elastic or plastic. We do so by computing a so-called trial stress $\boldsymbol{\sigma}^t$, which is the correct stress for an elastic step, but has to be corrected for a plastic step. The calculation of the strain tensor and its projection onto the FE-space is explained in the previous section.

$$(3.5.1) \qquad \boldsymbol{\sigma}_n^t = \left(\lambda(\theta_n, p_n) \mathrm{tr}(\boldsymbol{\epsilon}(\boldsymbol{u}_n)) - K(\theta_n, p_n) \frac{\rho_0 - \rho(\theta_n, p_n)}{\rho(\theta_n, p_n)} \right) \boldsymbol{I} +$$
$$+ 2\mu(\theta_n, p_n) \left(\boldsymbol{\epsilon}(\boldsymbol{u}_n) - \boldsymbol{\epsilon}_{n-1}^{cp} - \boldsymbol{\epsilon}_{n-1}^{up} \right)$$

With this trial stress we can evaluate the trial yield function.

$$(3.5.2) \qquad F_n^t = ||\boldsymbol{\sigma}_n^{t*} - \boldsymbol{X}_{n-1}^{cp}|| - \sqrt{\frac{2}{3}} \left(R_0(\theta_n, p_n) + R_{n-1}(s_{n-1}^{cp}, \theta_n, p_n) \right)$$

When the result is negative, then we are in an elastic time step and after a possible update of the TRIP tensor we are done with this time step. Otherwise we are in a plastic time step. Combinations of stress, back stress and radius of the yield surface which cause the yield function to be positive are physically not admissible, therefore we have to find a corrected stress $\boldsymbol{\sigma}_n^c$, a new back stress $\boldsymbol{X}_n^{cp}$ and a new radius of the yield surface R_n, so that the yield function using these values is zero. The basic idea of the algorithm of Simo and Hughes is to use the yield condition for the calculation of the plastic multiplier, γ_n, whereby the update of all other plastic quantities is possible. In our noticeably more complicated case with non-linear hardening rules and the coupling with transformation induced plasticity we can, with some effort, follow this basic idea.

Before we will explain this in detail, we introduce for convenience the so-called effective stress for classical plasticity, $\boldsymbol{\xi}^{cp}$, and for transformation induced plasticity, $\boldsymbol{\xi}^{up}$, and their trial versions, which are defined as follows:

$$(3.5.3a) \qquad \boldsymbol{\xi}_n^{cp} := \boldsymbol{\sigma}_n^c - \boldsymbol{X}_n^{cp} \qquad\qquad \boldsymbol{\xi}_n^{up} := \boldsymbol{\sigma}_n^c - \boldsymbol{X}_n^{up}$$
$$(3.5.3b) \qquad \boldsymbol{\xi}_n^{cp,t} := \boldsymbol{\sigma}_n^t - \boldsymbol{X}_{n-1}^{cp} \qquad\qquad \boldsymbol{\xi}_n^{up,t} := \boldsymbol{\sigma}_n^t - \boldsymbol{X}_{n-1}^{up}$$

What we aim to do is to use the yield condition

$$(3.5.4) \qquad 0 = ||\boldsymbol{\xi}_n^{cp}|| - \sqrt{\frac{2}{3}} \left(R_0(\theta_n, p_n) + R_n(s_n^{cp}, \theta_n, p_n) \right)$$

for the computation of the plastic multiplier. Therefore we have to find expressions for all quantities in equation (3.5.4) wherein the plastic multiplier is the only unknown. Then the equation can be solved numerically.

We will start with the more difficult part of finding an expression for $||\boldsymbol{\xi}_n^{cp}||$ which depends only on γ_n. The corrected stress can be defined and expressed as follows, using the equations for the trial stress (3.5.1) and the definitions for the plasticity and TRIP tensors in Box 2.4.3.

$$(3.5.5a) \qquad \boldsymbol{\sigma}_n^{c*} = 2\mu(\boldsymbol{\epsilon}^*(\boldsymbol{u}_n) - \boldsymbol{\epsilon}_n^{cp} - \boldsymbol{\epsilon}_n^{up})$$

$$(3.5.5b) \qquad = 2\mu(\boldsymbol{\epsilon}^*(\boldsymbol{u}_n) - \boldsymbol{\epsilon}_{n-1}^{cp} - \boldsymbol{\epsilon}_{n-1}^{up})$$
$$- 2\mu\tau_n(\boldsymbol{\epsilon}_n^{cp} - \boldsymbol{\epsilon}_{n-1}^{cp}) - 2\mu\tau_n(\boldsymbol{\epsilon}_n^{up} - \boldsymbol{\epsilon}_{n-1}^{up})$$

$$(3.5.5c) \qquad = \boldsymbol{\sigma}_n^{t*} - 2\mu\tau_n\gamma_n\frac{\boldsymbol{\xi}_n^{cp}}{||\boldsymbol{\xi}_n^{cp}||} - 3\mu\tau_n\kappa\boldsymbol{\xi}_n^{up}\sum_i^m \dot{\Phi}_i(p_i)\dot{p}_i$$

Now we look at the back stresses, using again the trick $\boldsymbol{X}_n = \boldsymbol{X}_{n-1} + \tau_n\dot{\boldsymbol{X}}$ and with the equation for the back-stresses in Box 2.4.3 (Prager model) we obtain:

$$(3.5.6a) \qquad \boldsymbol{X}_n^{cp} = \boldsymbol{X}_{n-1}^{cp} + \tau_n\frac{2}{3}c_{cp}\gamma_n\frac{\boldsymbol{\xi}_n^{cp}}{||\boldsymbol{\xi}_n^{cp}||} + \tau_n c_{int}\kappa\boldsymbol{\xi}_n^{up}\sum_i^m \dot{\Phi}_i(p_i)\dot{p}_i$$

$$(3.5.6b) \qquad \boldsymbol{X}_n^{up} = \boldsymbol{X}_{n-1}^{up} + \tau_n\frac{2}{3}c_{int}\gamma_n\frac{\boldsymbol{\xi}_n^{cp}}{||\boldsymbol{\xi}_n^{cp}||} + \tau_n c_{up}\kappa\boldsymbol{\xi}_n^{up}\sum_i^m \dot{\Phi}_i(p_i)\dot{p}_i$$

Inserting the above equations (3.5.5) and (3.5.6) into (3.5.3a) we gain expressions for the effective stresses.

$$(3.5.7a) \qquad \boldsymbol{\xi}_n^{cp} = \boldsymbol{\xi}_n^{cp,t} - (2\mu + \frac{2}{3}c_{cp})\tau_n\gamma_n\frac{\boldsymbol{\xi}_n^{cp}}{||\boldsymbol{\xi}_n^{cp}||} - (3\mu + c_{int})\tau_n\kappa\left(\sum_i^m \dot{\Phi}_i(p_i)\dot{p}_i\right)\boldsymbol{\xi}_n^{up}$$

$$(3.5.7b) \qquad \boldsymbol{\xi}_n^{up} = \boldsymbol{\xi}_n^{up,t} - (2\mu + \frac{2}{3}c_{int})\tau_n\gamma_n\frac{\boldsymbol{\xi}_n^{cp}}{||\boldsymbol{\xi}_n^{cp}||} - (3\mu + c_{up})\tau_n\kappa\left(\sum_i^m \dot{\Phi}_i(p_i)\dot{p}_i\right)\boldsymbol{\xi}_n^{up}$$

The next step is to solve (3.5.7b) for $\boldsymbol{\xi}_n^{up}$ and insert it into (3.5.7a). For convenience we introduce the abbreviation: $d := \frac{(3\mu+c_{int})\tau_n\kappa\left(\sum_i^m \dot{\Phi}_i(p_i)\dot{p}_i\right)}{1+(3\mu+c_{up})\tau_n\kappa\left(\sum_i^m \dot{\Phi}_i(p_i)\dot{p}_i\right)}$

$$(3.5.8a) \qquad \boldsymbol{\xi}_n^{up} = \frac{1}{1 + (3\mu + c_{up})\tau_n\kappa\left(\sum_i^m \dot{\Phi}_i(p_i)\dot{p}_i\right)}\left(\boldsymbol{\xi}_n^{up,t} - (2\mu + \frac{2}{3}c_{int})\tau_n\gamma_n\frac{\boldsymbol{\xi}_n^{cp}}{||\boldsymbol{\xi}_n^{cp}||}\right)$$

$$(3.5.8b) \qquad \boldsymbol{\xi}_n^{cp} = \boldsymbol{\xi}_n^{cp,t} - d\boldsymbol{\xi}_n^{up,t} - (2\mu + \frac{2}{3}c_{cp} - d(2\mu + \frac{2}{3}c_{int}))\tau_n\gamma_n\frac{\boldsymbol{\xi}_n^{cp}}{||\boldsymbol{\xi}_n^{cp}||}$$

We rearrange (3.5.8b) and look at the norm of the resulting equation.

$$(3.5.9) \qquad ||\boldsymbol{\xi}_n^{cp}|| \cdot \left| 1 + (2\mu + \frac{2}{3}c_{cp} - d(2\mu + \frac{2}{3}c_{int}))\tau_n\gamma_n \frac{1}{||\boldsymbol{\xi}_n^{cp}||} \right| = ||\boldsymbol{\xi}_n^{cp,t} - d\boldsymbol{\xi}_n^{up,t}||$$

Since the inner term is non-negative, taking the absolute value in the middle of the above equation can be dropped (as can easily be seen, considering that yields: $c_{int}^2 \leq c_{up}c_{cp}$). In this way we obtain the searched form of $||\boldsymbol{\xi}_n^{cp}||$ where the plastic multiplier, γ_n, is the only unknown.

$$(3.5.10) \qquad ||\boldsymbol{\xi}_n^{cp}|| = ||\boldsymbol{\xi}_n^{cp,t} - d\boldsymbol{\xi}_n^{up,t}|| - (2\mu + \frac{2}{3}c_{cp} - d(2\mu + \frac{2}{3}c_{int}))\tau_n\gamma_n$$

Remark. For later use it is worth noting that $\frac{\boldsymbol{\xi}_n^{cp}}{||\boldsymbol{\xi}_n^{cp}||} = \frac{\boldsymbol{\xi}_n^{cp,t} - d\boldsymbol{\xi}_n^{up,t}}{||\boldsymbol{\xi}_n^{cp,t} - d\boldsymbol{\xi}_n^{up,t}||}$ holds, which can easily be verified using the above equations.

Now we specify an expression for the increase of the yield surface, R, which depends only on the plastic multiplier.

$$(3.5.11a) \qquad R_n = \sum_{i=1}^{m} p_i c_i (s_n^{cp})^{m_i}$$

$$(3.5.11b) \qquad = \sum_{i=1}^{m} p_i c_i \left(s_{n-1}^{cp} + \tau_n \sqrt{\frac{2}{3}}\gamma_n \right)^{m_i}$$

When we insert equations (3.5.10) and (3.5.11b) into the yield condition (3.5.4), we obtain the equation which we can use for the calculation of the plastic multiplier γ_n via numerical root search, e.g. by Newton method or by bisection method.

$$(3.5.12) \qquad 0 = ||\boldsymbol{\xi}_n^{cp,t} - d\boldsymbol{\xi}_n^{up,t}|| - (2\mu + \frac{2}{3}c_{cp} - d(2\mu + \frac{2}{3}c_{int}))\tau_n\gamma_n +$$
$$- \sqrt{\frac{2}{3}} \left(R_0 + \sum_{i=1}^{m} p_i c_i \left(s_{n-1}^{cp} + \tau_n \sqrt{\frac{2}{3}}\gamma_n \right)^{m_i} \right)$$

After the derivation of the calculation of the new plastic multiplier we can now write down the algorithm for the calculation of classical plasticity. For this we assemble some of the above equations and for shortening the formulas we use $\frac{\boldsymbol{\xi}_n^{cp,t} - d\boldsymbol{\xi}_n^{up,t}}{||\boldsymbol{\xi}_n^{cp,t} - d\boldsymbol{\xi}_n^{up,t}||} = \frac{\boldsymbol{\xi}_n^{cp}}{||\boldsymbol{\xi}_n^{cp}||}$.

3.5.2. Algorithm for plasticity with isotropic and kinematic hardening:

- *Calculate trial stress*

$$(3.5.13) \qquad \boldsymbol{\sigma}_n^{t*} = 2\mu\left(\boldsymbol{\epsilon}(\boldsymbol{u}_n) - \boldsymbol{\epsilon}_{n-1}^{cp} - \boldsymbol{\epsilon}_{n-1}^{up}\right)$$

$$(3.5.14) \qquad \boldsymbol{\xi}_n^{cp,t} = \boldsymbol{\sigma}_n^{t*} - \boldsymbol{X}_{n-1}^{cp}$$

$$(3.5.15) \qquad \boldsymbol{\xi}_n^{up,t} = \boldsymbol{\sigma}_n^{t*} - \boldsymbol{X}_{n-1}^{up}$$

- *Evaluate yield function, using trial effective stress and R_{n-1}*

$$F_n^t = ||\boldsymbol{\xi}_n^{cp,t}|| - \sqrt{\frac{2}{3}}\left(R_0(\theta_n,p_n) + R_{n-1}(s_{n-1}^{cp},\theta_n,p_n)\right)$$

- **If** $F < 0$: *elastic time step*

  - *Plastic multiplier:* $\lambda = 0$

- **Else**: *plastic time step*

  - *Compute plastic multiplier numerically*
    for brevity we set: $d := \dfrac{\left(3\mu+\frac{3}{2}c_{int}\right)\tau_n\kappa\left(\sum_{i=1}^m \dot{\Phi}_i(p_i)\dot{p}_i\right)}{1+\left(3\mu+\frac{3}{2}c_{up}\right)\tau_n\kappa\left(\sum_{i=1}^m \dot{\Phi}_i(p_i)\dot{p}_i\right)}$

$$0 = ||\boldsymbol{\xi}_n^{cp}|| - \sqrt{\frac{2}{3}}(R_0 + R_n)$$

$$= ||\boldsymbol{\xi}_n^{cp,t} - d\boldsymbol{\xi}_n^{up,t}|| - (2\mu + c_{cp} - d(2\mu + c_{int}))\tau_n\gamma_n +$$

$$- \sqrt{\frac{2}{3}}\left(R_0 + \sum_{i=1}^m p_i c_i \left(s_{n-1}^{cp} + \tau_n\sqrt{\frac{2}{3}}\gamma_n\right)^{m_i}\right)$$

- *Update plastic and TRIP quantities, if necessary*

$$\boldsymbol{\xi}_n^{cp} = (\boldsymbol{\xi}_n^{cp,t} - d\boldsymbol{\xi}_n^{up,t}) - (2\mu + c_{cp} - d(2\mu + c_{int}))\tau_n\gamma_n \frac{\boldsymbol{\xi}_n^{cp}}{||\boldsymbol{\xi}_n^{cp}||}$$

$$\boldsymbol{\xi}_n^{up} = \frac{\left(\boldsymbol{\xi}_n^{up,t} - (2\mu + c_{int})\tau_n\gamma_n \frac{\boldsymbol{\xi}_n^{cp}}{||\boldsymbol{\xi}_n^{cp}||}\right)}{1 + \left(3\mu + \frac{3}{2}c_{up}\right)\tau_n\kappa\left(\sum_{i=1}^m \dot{\Phi}_i(p_i)\dot{p}_i\right)}$$

$$s_n^{cp} = s_{n-1}^{cp} + \tau_n \sqrt{\frac{2}{3}} \gamma_n$$

$$R_n = \sum_{i=1}^{m} p_i c_i(\theta) (s_n^{cp})^{m_i(\theta)}$$

$$\boldsymbol{\sigma}_n^{c*} = \boldsymbol{\sigma}_n^{t*} - 2\mu\tau_n\gamma_n \frac{\boldsymbol{\xi}_n^{cp}}{||\boldsymbol{\xi}_n^{cp}||} - 3\mu\tau_n\kappa\boldsymbol{\xi}_n^{up} \sum_{i=1}^{m} \dot{\Phi}_i(p_i)\dot{p}_i$$

$$\boldsymbol{X}_n^{cp} = \boldsymbol{X}_{n-1}^{cp} + \tau_n \frac{2}{3} c_{cp} \gamma_n \frac{\boldsymbol{\xi}_n^{cp}}{||\boldsymbol{\xi}_n^{cp}||} + \tau_n c_{int}\kappa\boldsymbol{\xi}_n^{up} \sum_{i=1}^{m} \dot{\Phi}_i(p_i)\dot{p}_i$$

$$\boldsymbol{X}_n^{up} = \boldsymbol{X}_{n-1}^{up} + \tau_n \frac{2}{3} c_{int} \gamma_n \frac{\boldsymbol{\xi}_n^{cp}}{||\boldsymbol{\xi}_n^{cp}||} + \tau_n c_{up}\kappa\boldsymbol{\xi}_n^{up} \sum_{i=1}^{m} \dot{\Phi}_i(p_i)\dot{p}_i$$

$$\epsilon_n^{cp} = \epsilon_{n-1}^{cp} + \tau_n\gamma_n \frac{\boldsymbol{\xi}_n^{cp}}{||\boldsymbol{\xi}_n^{cp}||}$$

$$\epsilon_n^{up} = \epsilon_{n-1}^{up} + \frac{3}{2}\tau_n\kappa\boldsymbol{\xi}_n^{up} \sum_{i=1}^{m} \dot{\Phi}_i(p_i)\dot{p}_i$$

Remark. The correction of the stress tensor by this algorithm is called *closest point projection*: The trial stress tensor lies outside the allowed stress space and produces a positive evaluation of the yield function. The corrected stress tensor is projected on the boundary of the allowed stress space. This is demonstrated for the (simpler) case of plasticity with isotropic hardening in Figures 3.1 and 3.2, the stress space grows simultaneously.

Remark. During the phase transformation there can be present transformation induced plasticity. Due to the interaction between classical plasticity and transformation induced plasticity, an update of the plastic quantities might be necessary even in elastic time steps.

This algorithm is one way to conduct step 4 in the overall scheme.

3.6 Realisation of periodic and symmetric boundary conditions

In this section we will explain how two non-standard boundary conditions are included into the implementation. The first subsection will deal with periodic boundary conditions and the second will deal with symmetric boundary conditions. In both cases the changes

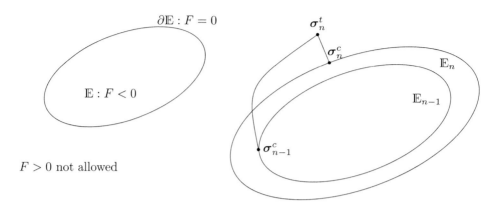

Figure 3.1: Stress space and the yield function

Figure 3.2: Closest point projection of trial stress for plasticity with isotropic hardening

in the calculation of scalar, vectorial and tensorial quantities are described. When we integrate these changes into our overall solution scheme in Box 3.2.1, only the calculation of temperature (scalar quantity), deformation (vectorial quantity) and the projection of the strain tensor (tensorial quantity) are affected. All other quantities, like phases, stress or the plastic quantities, are automatically periodic or symmetric when they are calculated from the corresponding periodic or symmetric temperature, deformation and strain. Therefore we will keep the following explanations abstract.

3.6.1   Periodic boundary conditions

For symmetric problems it can be very efficient to use periodic boundary conditions. For example when a ring is quenched symmetrically (and no other asymmetric effects are present), then it is sufficient to simulate only a small cutout of the ring with periodic boundary conditions on both cutting sides, compare Figure 3.3(a). This reduces the computation time enormously and the behaviour of the complete ring can still easily be constructed via periodic continuation.

The idea in this realisation of periodic boundary conditions is to identify all degrees of freedom on one periodic boundary with the degrees of freedom on the other periodic boundary (identify pairwise degrees of freedom with coordinates $(r, 0, h)$ and (r, α, h)), compare Figure 3.3(b). By doing so, the number of degrees of freedom is reduced and periodic FE-spaces for scalar, vectorial and tensorial quantities can be defined. We will explain this in detail for the different cases.

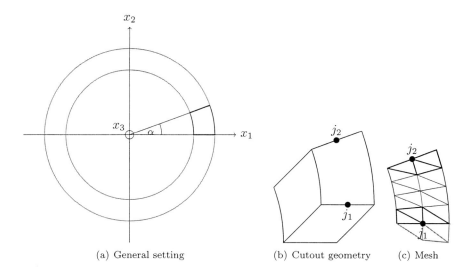

(a) General setting   (b) Cutout geometry   (c) Mesh

Figure 3.3: Schematically illustration of the application of periodic boundary conditions to a symmetric problem with a ring geometry.

Scalar quantities

As a first step we have to define the periodic boundary space, V_h^p, which will be used for the description of scalar quantities, like the temperature. In the periodic FE-space all Finite Elements, which belong to degrees of freedom, which do not lie on the periodic boundary, are the same as the Finite Elements in the usual FE-space V_h (defined in equation (3.2.1)). Those Finite Elements on the periodic boundary, which belong to two identified degrees of freedom are added to one Finite Element in the periodic FE-space: $\tilde{\varphi}_{j_1} = \varphi_{j_1} + \varphi_{j_2}$. The support of these Finite Elements spans on both sides of the periodic boundaries, compare Figure 3.3(c). In this way we define the periodic boundary space, V_h^p, of dimension $N - k$, where k is the number of degrees of freedom on one periodic boundary.

$$(3.6.1) \qquad\qquad V_h^p := \mathrm{span}(\tilde{\varphi}_1, \dots, \tilde{\varphi}_{N-k})$$

For the calculation of the scalar quantity the linear system of equations (corresponding to its discretised equation) is assembled using the usual FE-space V_h associated with homogeneous Neumann boundary conditions on the periodic boundaries. For the temperature this means, that the discretised heat equation in Box 3.3.1 is assembled, and on the periodic boundaries there is no heat flux in normal direction. Before the system of equations is solved the identification of the degrees of freedom is realised using the identification operator D^s. We denote by A the system matrix, by F the right hand side and by x^s the sought scalar quantity and obtain:

$$(3.6.2) \qquad\qquad (D^s)^T A D^s\, x^s = (D^s)^T F$$

In the following we will specify matrix D^s as an example for the two identified degrees of freedom j_1 and j_2.

$$(3.6.3) \qquad D^s = \begin{pmatrix} 1 & & & & & & \\ & \ddots & & & & & \\ & & 1 & & & & \\ & & 1 & 0 & & & \\ & & & 1 & & & \\ & & & & \ddots & & \\ & & & & & 1 \end{pmatrix}$$
$$ \begin{array}{cc} \uparrow & \uparrow \\ j_1 & j_2 \end{array}$$

The matrix for all k identified degrees of freedom is obtained analogously.

Vectorial quantities

In the case of a vectorial quantity we proceed as for the scalar quantity, but the definition of the periodic FE-space and the matrix of the identification operator is more complicated. On the periodic boundary the first two components of vector $\boldsymbol{x}^v(j_1)$ rotated by the angle α must be the same as the first two components of vector $\boldsymbol{x}^v(j_2)$. The third component of both vectors has to be equal.

$$(3.6.4) \qquad \begin{pmatrix} x_1^v \\ x_2^v \\ x_3^v \end{pmatrix}(j_1) = \begin{pmatrix} \cos(\alpha) & \sin(\alpha) & 0 \\ -\sin(\alpha) & \cos(\alpha) & 0 \\ 0 & 0 & 1 \end{pmatrix} \begin{pmatrix} x_1^v \\ x_2^v \\ x_3^v \end{pmatrix}(j_2)$$

In the periodic FE-space for vectors, V_h^{3p}, all Finite Elements, which belong to degrees of freedom, which do not lie on the periodic boundary, are the same as the Finite Elements in the usual FE-space V_h^3 (defined in equation (3.2.2)). Those Finite Elements on the periodic boundary, which belong to two identified degrees of freedom belong to one Finite Element in the periodic FE-space. Equation (3.6.4) motivates the form of the Finite Elements on the periodic boundary:

$$\tilde{\phi}_{j1}^1 := \phi_{j1}^1 + \cos(\alpha)\phi_{j2}^1 + \sin(\alpha)\phi_{j2}^2$$
$$\tilde{\phi}_{j1}^2 := \phi_{j1}^2 - \sin(\alpha)\phi_{j2}^1 + \cos(\alpha)\phi_{j2}^2$$
$$\tilde{\phi}_{j1}^3 := \phi_{j1}^3 + \phi_{j2}^3$$

Now we can define the periodic boundary space for vectorial quantities, V_h^{3p}, of dimension $3(N-k)$, where k is the number of degrees of freedom on one periodic boundary.

$$(3.6.5) \qquad\qquad V_h^{3p} := \mathrm{span}(\tilde{\phi}_1, \ldots, \tilde{\phi}_{3(N-k)})$$

Analogously to the procedure for scalar quantities, the system of linear equations is assembled using the usual FE-space V_h^3 associated with homogeneous Neumann boundary conditions on the periodic boundary. For the deformation this means that the linear system of equations given by the discretised deformation equation in Box 3.3.2 is assembled normally with no external forces applied in normal direction at the periodic boundaries. Before the system of equations is solved the identification of the degrees of freedom is realised using the identification operator D^v. As the system matrix $A^{3\times3}$ is a block matrix of three times three quadratic blocks of dimension N, the matrix of the identification operator D^v, is also a block matrix. We denote by F^3 the right hand side of the linear system of equations and by x^v the sought vectorial quantity and obtain:

$$(3.6.6) \qquad\qquad (D^v)^T A^{3\times3} D^v x^v = (D^v)^T F^3$$

In the following we will specify the matrix D^v as an example for the two identified degrees of freedom j_1 and j_2.

$D^v =$

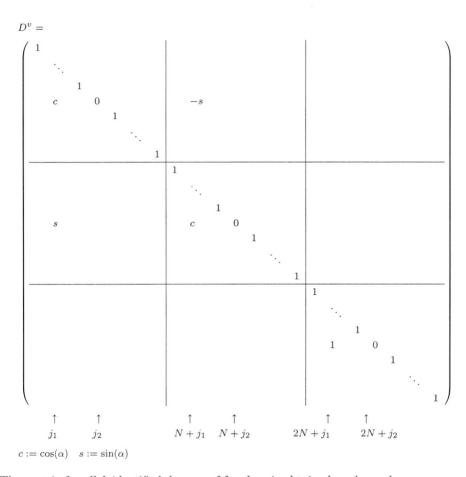

$c := \cos(\alpha) \quad s := \sin(\alpha)$

The matrix for all k identified degrees of freedom is obtained analogously.

Tensorial quantities

In the case of a tensorial quantity we proceed as for the tensorial quantity, but the definition of the periodic FE-space and the matrix of change of Basis is again more complicated. In the periodic FE-space for tensorial quantities, V_h^{6p}, all Finite Elements, which belong to degrees of freedom, which do not lie on the periodic boundary, are the same as the Finite Elements in the usual FE-space V_h^6 (defined in equation (3.2.3)). Those Finite Elements on the periodic boundary, which belong to two identified degrees of freedom belong to one Finite Element in the periodic FE-space and are defined as follows (using the rotation matrix R in equation (3.6.4)):

$$\tilde{\phi}_{j1}^{11} := \phi_{j1}^{11} + R^T \phi_{j2}^{11} R$$

$$= \phi_{j1}^{11} + \begin{pmatrix} \cos^2(\alpha) & \sin(\alpha)\cos(\alpha) & 0 \\ \sin(\alpha)\cos(\alpha) & \sin^2(\alpha) & 0 \\ 0 & 0 & 0 \end{pmatrix}_{j2}$$

(3.6.7a)
$$= \phi_{j1}^{11} + \cos^2(\alpha)\phi_{j2}^{11} + \sqrt{2}\sin(\alpha)\cos(\alpha)\phi_{j2}^{12} + \sin^2(\alpha)\phi_{j2}^{22}$$

$$\tilde{\phi}_{j1}^{22} := \phi_{j1}^{22} + R^T \phi_{j2}^{22} R$$

(3.6.7b)
$$= \phi_{j1}^{22} + \sin^2(\alpha)\phi_{j2}^{11} - \sqrt{2}\sin(\alpha)\cos(\alpha)\phi_{j2}^{12} + \cos^2(\alpha)\phi_{j2}^{22}$$

$$\tilde{\phi}_{j1}^{12} := \phi_{j1}^{12} + R^T \phi_{j2}^{12} R$$

(3.6.7c)
$$= \phi_{j1}^{12} - \sqrt{2}\sin(\alpha)\cos(\alpha)\phi_{j2}^{11} + (\cos^2(\alpha) - \sin^2(\alpha))\phi_{j2}^{12} +$$
$$+ \sqrt{2}\sin(\alpha)\cos(\alpha)\phi_{j2}^{22}$$

$$\tilde{\phi}_{j1}^{13} := \phi_{j1}^{13} + R^T \phi_{j2}^{13} R$$

(3.6.7d)
$$= \phi_{j1}^{13} + \cos(\alpha)\phi_{j2}^{13} + \sin(\alpha)\phi_{j2}^{23}$$

$$\tilde{\phi}_{j1}^{23} := \phi_{j1}^{23} + R^T \phi_{j2}^{23} R$$

(3.6.7e)
$$= \phi_{j1}^{23} - \sin(\alpha)\phi_{j2}^{13} + \cos(\alpha)\phi_{j2}^{23}$$

$$\tilde{\phi}_{j1}^{33} := \phi_{j1}^{33} + R^T \phi_{j2}^{33} R$$

(3.6.7f)
$$= \phi_{j1}^{33} + \phi_{j2}^{33}$$

In this way we define the periodic FE-space for tensorial quantities, V_h^{6p}, which is of dimension $6(N - k)$, where k is the number of degrees of freedom on one periodic boundary.

(3.6.8)
$$V_h^{6p} := \text{span}(\tilde{\phi}_1, \ldots, \tilde{\phi}_{6(N-k)})$$

Analogously to the procedures for scalar and tensorial quantities, the system of linear equations is assembled using the usual FE-space V_h^6 associated with homogeneous Neumann boundary conditions on the periodic boundary. In the calculation of the projection of the strain tensor there is a fundamental change necessary. As explained in Section 3.4 this projection is normally calculated component wise, because there is no coupling between the components. This is no longer possible for periodic boundary conditions. So we define the system matrix to be a block matrix of six times six quadratic blocks of dimension N, where on the diagonal there is six times the primal system matrix. The right hand side is gained in appending the right hand side for all six components of the strain to a vector of length $6N$.

Before the system of equations is solved the identification of the degrees of freedom is realised using the identification operator D^t. As the system matrix $A^{6\times6}$ is a block matrix, the matrix of the identification operator D^t is also a block matrix. We denote by F^6 the right hand side of the linear system of equations and by x^t the sought vectorial quantity and obtain:

(3.6.9)
$$(D^t)^T A^{6\times6} D^t\, x^t = (D^t)^T F^6$$

In the following we will specify matrix D^t for the two identified degrees of freedom j_1 and j_2. D^t is a quadratic block matrix of dimension $6N$. Due to spatial restrictions, we specify only the entries which deviate from the identity matrix.

$$
\begin{aligned}
(D^t)_{ii} &= 0 & & i = nN + j_2, n = 0, \ldots, 5 \\
(D^t)_{ij} &= \cos^2(\alpha) & & (i,j) \in \{(j_1, j_2), (3N + j_1, 3N + j_2)\} \\
(D^t)_{ij} &= \sqrt{2}\cos(\alpha)\sin(\alpha) & & (i,j) \in \{(N + j_1, j_2), (3N + j_1, N + j_2)\} \\
(D^t)_{ij} &= -\sqrt{2}\cos(\alpha)\sin(\alpha) & & (i,j) \in \{(j_1, N + j_2), (N + j_1, 3N + j_2)\} \\
(D^t)_{ij} &= \sin^2(\alpha) & & (i,j) \in \{(j_1, 3N + j_2), (3N + j_1, j_2)\} \\
(D^t)_{ij} &= \cos^2(\alpha) - \sin^2(\alpha) & & (i,j) \in \{(N + j_1, N + j_2)\} \\
(D^t)_{ij} &= \cos(\alpha) & & (i,j) \in \{(2N + j_1, 2N + j_2), (4N + j_1, 4N + j_2)\} \\
(D^t)_{ij} &= \sin(\alpha) & & (i,j) \in \{(4N + j_1, 2N + j_2)\} \\
(D^t)_{ij} &= -\sin(\alpha) & & (i,j) \in \{(2N + j_1, 4N + j_2)\} \\
(D^t)_{ij} &= 1 & & (i,j) \in \{(5N + j_1, 5N + j_2)\}
\end{aligned}
$$

The matrix for all k identified degrees of freedom is obtained analogously.

These are all changes necessary to our overall solution scheme in Box 3.2.1 for the application of periodic boundary conditions.

3.6.2   Symmetric boundary conditions

For problems which are symmetric to one, two or three axes the use of symmetric boundary conditions is very useful. The computational geometry can be reduced by clipping along these axes and so only a smaller cutout has to be simulated, compare Figure 3.4. This reduces the computing time considerably and the behaviour of the complete geometry

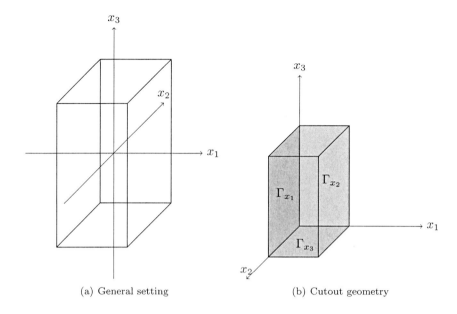

(a) General setting   (b) Cutout geometry

Figure 3.4: Schematic illustration of the application of symmetric boundary condition to an symmetric problem with a cuboid geometry.

can still easily be constructed via mirroring along the corresponding axes.

We will now explain how the symmetric boundary conditions are realised for the scalar, vectorial and tensorial quantities in our setting. This can be done relatively easily, as it is sufficient to apply (if necessary, component by component) homogeneous Dirichlet or Neumann boundary conditions.

From the scalar quantities only the temperature is affected. At the symmetric boundaries a homogeneous Neumann boundary condition is applied. This means that there is no heat flux in normal direction at these part of the cutout geometry, as it would be the case if the complete geometry would be considered.

For the calculation of the deformation, which is the only vectorial quantity, we apply homogeneous Dirichlet boundary conditions for the correspondent component at the symmetric boundaries:

$$(3.6.10a) \qquad\qquad u_1 = 0 \quad \text{on } \Gamma_{x_1}$$

$$(3.6.10b) \qquad\qquad u_2 = 0 \quad \text{on } \Gamma_{x_2}$$

$$(3.6.10c) \qquad\qquad u_3 = 0 \quad \text{on } \Gamma_{x_3}$$

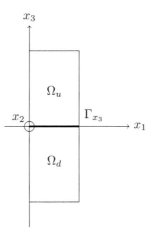

Figure 3.5: Schematic illustration of the application of symmetric boundary conditions concerning the strain tensor.

In the tensorial case only the projection of the strain tensor, as described in Section 3.4, is affected by the symmetric boundary conditions. The projection can still be done component by component but some additional boundary conditions have to be applied at the symmetric boundary. For some components, homogeneous Neumann and for others homogeneous Dirichlet boundary conditions have to be applied, respectively. We will use Γ_{x_3} as an example.

Let our domain Ω be the union of Ω_u and Ω_d, compare Figure 3.5. We define the mapping $C : \Omega_u \to \Omega_d$, which maps a point in Ω_u to its corresponding point in Ω_d, via:

$$(3.6.11) \qquad \boldsymbol{x} \in \Omega_u, \, \boldsymbol{x} = \begin{pmatrix} x_1 \\ x_2 \\ x_3 \end{pmatrix} \quad C(\boldsymbol{x}) := \begin{pmatrix} x_1 \\ x_2 \\ -x_3 \end{pmatrix}$$

Moreover we define the mapping $F : \Omega \to \mathbb{R}^3$, which maps a point in Ω to the deformation vector in this point, via:

$$(3.6.12) \qquad F(\boldsymbol{x}) := \begin{pmatrix} u_1 \\ u_2 \\ u_3 \end{pmatrix} (\boldsymbol{x})$$

From equations (3.6.11) and (3.6.12) and the symmetry assumption, it follows directly that it holds:

$$(3.6.13) \qquad F(C(\boldsymbol{x})) = \begin{pmatrix} u_1 \\ u_2 \\ u_3 \end{pmatrix} (C(\boldsymbol{x})) = \begin{pmatrix} u_1 \\ u_2 \\ -u_3 \end{pmatrix} (\boldsymbol{x}) \quad \Rightarrow u_3 \big|_{\Gamma_{x_3}} = 0$$

For the calculation of the strain tensor we need the derivatives of $\boldsymbol{u}$ and we want to compare the values of the Jacobian matrix of F in corresponding points in Ω_u and Ω_d.

$$(3.6.14a) \qquad DF(\boldsymbol{x}) = \begin{pmatrix} \frac{\partial u_1}{\partial x_1} & \frac{\partial u_1}{\partial x_2} & \frac{\partial u_1}{\partial x_3} \\ \frac{\partial u_2}{\partial x_1} & \frac{\partial u_2}{\partial x_2} & \frac{\partial u_2}{\partial x_3} \\ \frac{\partial u_3}{\partial x_1} & \frac{\partial u_3}{\partial x_2} & \frac{\partial u_3}{\partial x_3} \end{pmatrix}$$

$$DF(C(\boldsymbol{x})) = DF(C(\boldsymbol{x})) \cdot DC(\boldsymbol{x})$$

$$= \begin{pmatrix} \frac{\partial u_1}{\partial x_1} & \frac{\partial u_1}{\partial x_2} & \frac{\partial u_1}{\partial x_3} \\ \frac{\partial u_2}{\partial x_1} & \frac{\partial u_2}{\partial x_2} & \frac{\partial u_2}{\partial x_3} \\ -\frac{\partial u_3}{\partial x_1} & -\frac{\partial u_3}{\partial x_2} & -\frac{\partial u_3}{\partial x_3} \end{pmatrix} \cdot \begin{pmatrix} 1 & 0 & 0 \\ 0 & 1 & 0 \\ 0 & 0 & -1 \end{pmatrix}$$

$$(3.6.14b) \qquad = \begin{pmatrix} \frac{\partial u_1}{\partial x_1} & \frac{\partial u_1}{\partial x_2} & -\frac{\partial u_1}{\partial x_3} \\ \frac{\partial u_2}{\partial x_1} & \frac{\partial u_2}{\partial x_2} & -\frac{\partial u_2}{\partial x_3} \\ -\frac{\partial u_3}{\partial x_1} & -\frac{\partial u_3}{\partial x_2} & +\frac{\partial u_3}{\partial x_3} \end{pmatrix}$$

From the above equations it follows directly:

$$(3.6.15) \qquad \frac{\partial u_3}{\partial x_1}\bigg|_{\Gamma_{x_3}} = \frac{\partial u_3}{\partial x_2}\bigg|_{\Gamma_{x_3}} = \frac{\partial u_1}{\partial x_3}\bigg|_{\Gamma_{x_3}} = \frac{\partial u_2}{\partial x_3}\bigg|_{\Gamma_{x_3}} = 0$$

All other derivatives $\frac{\partial u_1}{\partial x_1}, \frac{\partial u_1}{\partial x_2}, \frac{\partial u_2}{\partial x_2}, \frac{\partial u_3}{\partial x_3}$ are the same in $\boldsymbol{x}$ and $C(\boldsymbol{x})$. With analogous considerations we obtain the following boundary conditions for the projection of the strain tensor on the symmetric boundaries.

$$(3.6.16a) \qquad \epsilon_{12} = \epsilon_{13} = 0 \quad \text{on } \Gamma_{x_1}$$

$$(3.6.16b) \qquad \epsilon_{12} = \epsilon_{23} = 0 \quad \text{on } \Gamma_{x_2}$$

$$(3.6.16c) \qquad \epsilon_{13} = \epsilon_{23} = 0 \quad \text{on } \Gamma_{x_3}$$

These are all changes necessary to our overall solution scheme in Box 3.2.1 for the application of symmetric boundary conditions.

3.7 Error indicators

As already mentioned before, we implemented our problem into the adaptive Finite Element toolbox ALBERTA, see [SS05]. As an open-source toolbox, ALBERTA provides an efficient implementation of modern numerical algorithms which offers the possibility to be adapted to scientific progress. Besides this, the main feature of ALBERTA is the adaptivity. The grid can be refined and coarsened according to user-specified *a posteriori error indicators*. Several refinement and coarsening strategies, like the maximum or the equi-distribution strategy are provided, as well as different overall strategies for the time discretisation of the problem. In this section we want to explain in detail, how the adaptive algorithm works and which a posteriori error indicators we use for adaptivity in time and space.

For time dependent problems not only the error caused by spatial discretisation but also the initial error and the error caused by time discretisation has to be considered. We split the complete error , η, into four components:

- η_0: initial error

- η_h: error of discretisation in space

- η_c: error caused by the mesh coarsening between two time steps

- η_τ: error of discretisation in time

For parabolic problems like the heat equation the error between the approximated solution u_N and the exact solution usually can be written like this, compare [SS05]:

$$||u(t_N) - u_N|| \leq \eta_0 + \max_{1 \leq n \leq N} \left(\eta_{\tau,n} + \left(\sum_{S \in \mathcal{T}_n} (\eta_{h,S}^q + \eta_{c,S}^q) \right)^{\frac{1}{q}} \right)$$

In this representation, the space discretisation error and the coarsening error are given by the sum over the local error indicators $\eta_{h,S}$ and $\eta_{c,S}$ on every element of the mesh $\mathcal{T}$. For an adaptive mesh, the entire error should be below, but close to a given tolerance tol. This is achieved in introducing portions Γ for the component errors and requiring that:

$$\Gamma_0 + \Gamma_\tau + \Gamma_h \leq 1, \quad \eta_0 \approx \Gamma_0 \text{tol}, \quad \eta_\tau \approx \Gamma_\tau \text{tol}, \quad \eta_h^q + \eta_c^q \approx (\Gamma_h \text{tol})^q$$

The calculation of these error components is then embedded into an algorithm which controls either only the time step size adjustment or both the time step size adjustment and the mesh adaption (refinement and coarsening).

As in our case we will control the time step size mainly by the heat equation but the mesh adaption by the deformation equation, we will now describe the algorithm for time step size adjustment only.

3.7.1. Strategy for time step size control:

- *start with old time step size, solve the problem and estimate the error*

- **While** $\eta_\tau > \alpha_1 \Gamma_\tau tol$:

    - *reduce time step size*

    - *solve again and estimate the error*

- **If** $\eta_\tau < \alpha_2 \Gamma_\tau tol$: *enlarge time step size*

where α_1, α_2 are parameters to enhance numerical stability.

Although we use the linear heat equation only for the time step size control we will specify time and spatial a posteriori error indicators:

$$
\begin{aligned}
\eta_{\tau,n}^2 =&\, C_0 ||\theta_n - \theta_{n-1}||_{L^2(\Omega)}^2 \\
\eta_{h,S}^2 =&\, C_1^2 h^4 \left\| \rho_0 c_e \frac{\theta_n - \theta_{n-1}}{\tau_n} - \nabla \kappa \nabla \theta_n - \rho_0 L \frac{p_n - p_{n-1}}{\tau_n} \right\|_{L^2(S)}^2 + \\
&\, + C_2^2 h^3 ||\kappa \nabla \theta_n \cdot n - \delta(\theta - \theta_{ext})||_{L^2(\partial S \cap \partial \Omega)}^2 + \\
&\, + C_3^2 h^3 ||[\kappa \nabla \theta]||_{L^2(\partial S \cap \Omega)}^2 \\
\eta_{c,S}^2 =&\, C_4^2 h^3 ||[\nabla \theta_n]||_{L^2(\Gamma_c)}^2
\end{aligned}
$$

(3.7.1)

Here $[\cdot]$ denotes the jump of a quantity over an interior edge/face and Γ_c denotes the edges/faces which are removed by coarsening in the n-th time step. The initial error is zero because our initial temperature $\theta_0 = $ const is an element of the Finite Element space.

Remark. It is worth noting that Erikkson and Johnson published a series of papers concerning adaptive Finite Element methods for parabolic problems. In [EJ91] they dealt with linear parabolic model problem and prove a priori and a posteriori estimates together with a reliable (concerning the L^2-error) and efficient adaptive method (efficient in the sense that the approximation error is for most time steps not essentially below the given tolerance).

For time step size control we do not use solely the heat equation but also consider contributions from phase change and classical plasticity. When there is a phase change, then also transformation induced plasticity occurs and both transformation induced and classical plasticity are processes which need small time steps (the semi-implicit algorithm we use for their calculation may need even smaller time steps). Therefore we use the following error indicator for time step size control.

3.7.2. Error indicator for time step size control:

$$\eta_{\tau,n} = c_0 ||\theta_n - \theta_{n-1}||^2_{L^2(\Omega)} + c_1 \max_{\boldsymbol{x} \in \Omega}(|p_n(\boldsymbol{x}) - p_{n-1}(\boldsymbol{x})|) +$$
$$+ c_2 \max_{\boldsymbol{x} \in \Omega} ||\boldsymbol{\epsilon}^{cp}_n(\boldsymbol{x}) - \boldsymbol{\epsilon}^{cp}_{n-1}(\boldsymbol{x})||$$

Here c_0, c_1 and c_2 are constants which can be used to ensure that the single effects are considered appropriately.

As already mentioned we use the deformation equation for the mesh refinement. In every time step we conduct the following algorithm for mesh refinement and coarsening.

3.7.3. Adaptive refinement strategy:

- *start with the old mesh, solve the problem and compute the local error indicators $\eta_{h,S}$ and $\eta_{c,S}$*

- **While** $\eta > tol$:

    - *mark elements for refinement and coarsening*

    - *adapt mesh*

    - *solve again and estimate the error*

The general idea of this algorithm is as follows: Compute a local error indicator on every Element S of the mesh $\mathcal{T}_n$. Use this indicator to decide whether this element is marked for refinement (or for coarsening) or not. Then apply a refinement or coarsening strategy provided by ALBERTA to adapt the mesh. We will use the equi-distribution strategy, which tries to keep the error of similar size on all mesh elements. In an ideal situation the error on all N elements of the mesh would be equal:

$$\eta = \left(\sum_{S \in \mathcal{T}_n} \eta_S^2 \right)^{\frac{1}{2}} = N^{\frac{1}{2}} \eta_S \overset{!}{=} tol \quad \Rightarrow \eta_S = \frac{tol}{N^{\frac{1}{2}}}$$

Therefore the equi-distribution strategy uses the following criterion for marking elements for refinement or coarsening

$$\eta_{h,S} > \theta \frac{tol}{N^{\frac{1}{2}}} \qquad\qquad\qquad \text{refine}$$

$$\eta_{h,S} + \eta_{c,S} \leq \theta_c \frac{tol}{N^{\frac{1}{2}}} \qquad\qquad \text{coarsen}$$

Here θ, θ_c are parameters to make the procedure more robust.

To apply this adaptive strategy to the quasi-static deformation equation, we need an a posteriori error estimators. In [Ver96] Verfürth proves a posteriori error estimation for a problem of linear elasticity. We adjust it to our setting and specify the error indicator in a simplified form in the box below.

3.7.4. Spatial error indicator (deformation equation):

$$\eta_{h,S} = c_1 h^2 \left\| - \operatorname{div}(\boldsymbol{\sigma}(\boldsymbol{u_n})) \right\|_{L^2(S)}^2 + c_2 h \left\| \boldsymbol{\sigma}(\boldsymbol{u_n}) \cdot \boldsymbol{n} - \boldsymbol{F}_{ext} \right\|_{L^2(\partial S \cap \partial \Omega)}^2 +$$
$$+ c_3 h \left\| \left[\boldsymbol{\sigma}(\boldsymbol{u_n}) \cdot \boldsymbol{n} \right] \right\|_{L^2(\partial S \cap \dot{\Omega})}^2$$
$$\eta_{c,S} = \left\| \left[\boldsymbol{\sigma}(\boldsymbol{u_n}) \cdot \boldsymbol{n} \right] \right\|_{L^2(\partial S \cap \Gamma_c)}^2$$

where $\boldsymbol{\sigma}(\boldsymbol{u_n})$ is computed via the following equation

$$\boldsymbol{\sigma}(\boldsymbol{u_n}) = 2\mu(\boldsymbol{\epsilon}(\boldsymbol{u_n}) - \boldsymbol{\epsilon}_{n-1}^{cp} - \boldsymbol{\epsilon}_{n-1}^{up}) + \left(\lambda \operatorname{tr}(\boldsymbol{\epsilon}(\boldsymbol{u_n})) - \left(K \frac{\rho_0 - \rho_n}{\rho_n} \right) \right) \boldsymbol{I}$$

and $[\cdot]$ denotes the jump of a quantity over an interior edge/face.

For the settings we are dealing with in the next chapter, where we will do simulations, we do not use adaptive mesh refinement but only the described time step size control. We are considering settings where the external conditions are invariable in time and do not vary strongly in space. Here the use of adaptive mesh refinement has no big advantages, unlike in settings with varying external conditions, e.g. in the simulation of laser welding with a moving heat source.

CHAPTER

4

Improved approximation of experiments via model comparison

In this chapter we want to use the mathematical models introduced in Section 2.4 and the numerical algorithms described in Chapter 3 for the simulation of quenching experiments of steel. There are several situations where experimental results and simulation results, which were obtained using wide-spread standard models, are in no good accordance. The aim is to reduce the deviation between experimental and simulation results by using different or more complex models.

The models under consideration are models for martensitic phase transformations, transformation induced plasticity and classical plasticity. In the comparison of the models we follow the philosophy of Mahnken [Mah04]. He distinguishes between the verification and the validation of a model. For the verification one dimensional experiments are used. These experiments are performed on a dilatometer type 805A (Bähr Thermo-analyse) and on a Gleeble 3500 thermomechanical simulator, which is a combination of a tension-compression testing machine and a quenching dilatometer. In the following these experiments will be referred to as dilatometer- and Gleeble tests. Using these one dimensional tests, it is checked whether the model is capable to describe the material behaviour at all, and whether the material parameters needed by the model can be identified. The next step is the validation of the model. In order to do this the model is used to simulate three dimensional experiments of work-pieces and it is checked whether the model describes the material behaviour well even in this more complex case.

In some cases not all experimental data necessary for the verification and validation of a model are available. For the verification we try to derive the necessary model parameters from comparable materials or, if this is not possible, we vary the parameters. Where validation experiments are missing, we compare different models (possibly with parameter variations) to investigate their influence on the simulation result.

There is still a noticeable problem with the validation part. Experimental results, obtained in the SFB 570 during the last six month, indicate, that the martensitic transformation in 100Cr6 steel has two characteristics which are not included into the classical martensite models. These characteristics are an isothermal martensitic transformation and an austenitic stabilisation. As the new results were available only recently and due to their complexity, the integration of these characteristics into our model remains future work.

In the first section of this chapter, we will discuss in more detail effects, which are not included in the modelling, but which could cause qualitative differences between experimental measurements and simulation results. This will, beside others, also include the new results on martensitic transformation already mentioned in the previous paragraph.

In the following three sections, different models will be compared in order to improve the accordance of simulated and experimental data. A tabular overview of the models and the experiments can be found in Table 4.1, an overview in text form is given below.

In the second section we will deal with the verification and validation of several models for the martensitic phase transformation and transformation induced plasticity. Concerning the transformation induced plasticity we will investigate the influence of different saturation functions. This work was finished before the new characteristics of martensitic transformation in 100Cr6 steel were discovered. The results obtained in this section remain valid, although they are restricted to this special setting (especially to this cooling rate) and no generalisations are possible.

In the third section we will do some test-simulations which will reassure us, that we can neglect the inertia term in the deformation equation for the settings we are dealing with.

The last section will deal with different hardening models for classical plasticity, the Ramberg-Osgood and Prager model and the Armstrong-Fredrick model, which both in-

4.2	Martensitic phase transformation		
	Models	Verification	Validation
	Koistinen-Marburger Leblond-Devaux Yu	dilatometer and Gleeble tests	moderately quenched conical rings
	Transformation induced plasticity: Saturation function		
	Models	Verification	Validation
	Tanaka Leblond Denis-Desalos Wolff	—	moderately quenched conical rings
4.3	Inertia term		
	Models	Verification	Validation
	standard	oscillating beam	moderately quenched conical rings
4.4	Classical plasticity		
	Models	Verification	Validation
	Ramberg-Osgood Prager Armstrong-Frederick	isothermal, cyclic Gleeble tests	quickly quenched conical rings, asymmetrically quenched rings

Table 4.1: Tabular overview of model comparisons in chapter four.

clude a coupling with the transformation induced plasticity. The model parameters of the Ramberg-Osgood and Prager model are available within the SFB 570. For the Armstrong-Fredrick model we derive an optimisation procedure for parameter identification from experimental data (this was done together with M. Wolff, University of Bremen). There are comparative simulations done, which show the different behaviour of both models, the effect and the extent of the coupling between classical plasticity and transformation induced plasticity. Finally a simulation of an asymmetrically quenched ring shows qualitative accordance with experimental measurements.

As already mentioned, all simulations are carried out in ALBERTA, an open source Finite Element toolbox, see [SS05]. As long as it is not explicitly indicated we use the material data set which was obtained in the SFB 570, see [ADF+08a, ADF+08b]. We are exclusively considering 100Cr6 steel (SAE 52100).

4.1 Preliminaries on interpretation of experimental data

As already said in the introductory part of this chapter, we want to compare simulations, which use different models, to experimental results. This approach clearly holds some difficulties as in the experiment there occur processes, which are not included in the modelling. In many cases one cannot be sure whether these processes influence the experimental results and if so to what extent. There are different reasons why these effects are not included in the modelling (yet). Some processes are considered to be negligible, while others are not included yet because there is missing either a model or experimental data, needed for the determination of model parameters. Within the SFB 570 there is work on progress to integrate the following processes into the modelling.

There are good reasons to believe that for the simulation of the heat treatment of work-pieces, it is important to include the heating-up process into the simulation. During this time there occur creep and transformation induced plasticity which both have a considerable influence on the work-piece. For both processes suitable models are known but the experimental data for the determination of model parameters is available only recently. Therefore the simulation of the heating-up of the work-piece remains future work.

A more difficult task arises from the recently found characteristics of the martensitic transformation. Results found by Hunkel [Hun] show, that there are two characteristics, which are not included into the standard models for martensitic transformation: isothermal martensitic transformation and austenitic stabilisation. These two characteristics have different effects on the transformation depending on the way of quenching. In continuous cooling processes a slower quenching leads to more isothermal transformation and to more austenitic stabilisation resulting in more rest austenite and the end of the quenching process compared to a quicker quenching. In cooling processes with isothermal holding there is at the beginning much isothermal transformation and towards the end much austenitic stabilisation. The amount of rest austenite at the end of the quenching process is most probably dependent on the duration of the holding. These processes are at the moment subject to experimental investigations. After a literature search it seems that only a few model approaches for single aspects of the above explained behaviour could be found, e.g. [Mag68] and [Sch83]. To the author's knowledge there exist no model for the combined effects of an-isothermal and isothermal martensitic transformation and austenitic stabilisation. Since the martensitic transformations in steel seems to be a well examined subject, it is especially surprising that the isothermal transformation and the austenitic stabilisation are hardly dealt with in the literature. The experimental investigation and the modelling of theses phenomena will be subject of future work within the SFB 570.

There are also processes occurring in experiments which can not be included in the modelling in the near future. An example for this category is the problem of Eigenstresses due to the machining process. When a ring is clamped during its production using a three jaw chuck, Eigenstresses are created. During the heat treatment these Eigenstresses are released and in the final distortion a triangularity can be seen. The modelling of this process is not advanced enough to include it into the modelling soon.

These are some of the reasons which complicate comparisons between experiments and simulations. The biggest effect is surely caused by the newly discovered characteristics of martensitic transformation, but also the other mentioned processes can have an considerable influence depending on the considered setting.

4.2 Martensitic transformation and TRIP

In this section we will compare three models for martensitic phase transformation and also we will vary the saturation function of transformation induced plasticity. This work is a continuation of the comparison of models for martensitic transformation and transformation induced plasticity done in [SFHW09, WBDH08]. The parameter optimisation for the martensite models is repeated using different experimental data, the current material data set of the SFB 570 (see [ADF+08a, ADF+08b]) and a different optimisation strategy.

For the optimisation of the martensite models we use dilatometer tests and stress free Gleeble tests (the latter were also used in [ADF+08b] for the fitting of the Koistinen-Marburger model). All tests have a continuous cooling but the dilatometer and Gleeble tests differ in their cooling rate which will cause different optimisation results. With the mentioned new results on martensitic transformation, this comparison between different models is only valid for the considered quenching situation. A generalisation to other settings is not possible.

The experimental data in [SFHW09] was not used, because the now chosen data is more reliable and shows less scattering of important quantities like martensite start temperature.

From other experiments, whose austenitising time is not in accordance with the SFB 570 material data set, we learned that the stress-dependence of the martensite start temperature is less then it was assumed in earlier works, e. g. [WBDH08]. Therefore we drop stress-dependency in phase transformation, martensite start temperature and the Greenwood-Johnson parameter of transformation induced plasticity until more reliable experimental data is available for investigation.

This section is organised as follows: In Subsection 4.2.1 we will repeat all used models for convenience. In Subsection 4.2.2 we will explain in detail how the model parameters for the phase transformation are gained via an optimisation procedure. Thereafter we will conduct the validation of the models via the three-dimensional simulation of the quenching of a conical ring in comparison to measured data, see Subsection 4.2.3.

4.2.1 Overview of tested models

In the following we will compare three different models for martensitic phase transformation as well as four proposals for the saturation function of transformation induced plasticity. In this subsection we will repeat these models from Subsections 2.2 and 2.3.4 for the reader's convenience.

Martensitic transformation

As already mentioned before, the experimental data available is difficult to interpret in means of, whether stress-dependent transformation behaviour plays an important role for 100Cr6 steel or not. In lack of reliable data for fitting stress-dependent model parameters, we deal with non-stress-dependent phase transformation models only. The models under consideration are: Koistinen-Marburger, Leblond-Devaux and Yu

$$(4.2.1\text{a}) \qquad \text{Koistinen-Marburger:} \qquad p = 1 - \exp^{-\frac{\theta_{ms} - \theta}{\theta_{m0}}}$$

$$(4.2.1\text{b}) \qquad \text{Leblond-Devaux:} \qquad \dot{p}(t) = \max\left(0, (\bar{p}(t) - p(t))\mu\right)$$

$$(4.2.1\text{c}) \qquad \text{Yu:} \qquad p(\theta) = \frac{\theta_{ms} - \theta}{\theta_{ms} - \theta - \beta(\theta_{mf} - \theta)}$$

Here θ_{ms} is the martensite start temperature and θ_{mf} is the martensite finish temperature. For the Leblond-Devaux model we denote by $\bar{p}(t)$ the maximal possible phase fraction given by the Koistinen-Marburger equation at time t and temperature θ. Each of the three models has one model parameter, which has to be determined via an optimisation procedure: θ_{m0}, μ and β.

The martensite start temperature of 100Cr6 steel was determined in [ADF+08a, ADF+08b]: $\theta_{ms} = 211°C$. We choose the martensite finish temperature to be $\theta_{mf} = -174°C$, because measurements at specimens quenched to this temperature had a fraction of rest-austenite of only 3%.

Transformation induced plasticity

In the literature there are several proposals for the saturation function Φ in the model for transformation induced plasticity. In [WBDH08] there were Gleeble-tests in tension and compression evaluated and a new saturation function by Wolff was presented.

The Gleeble-testing device is able to perform tests in tension or compression, whereas temperature, length, diameter and stress are measured. The tests are conducted with small cylinders and the following assumptions are used: spatial homogeneity and volume conservation of transformation induced plasticity. For the test used in [WBDH08] there was also the absence of classical plasticity assumed, see [DLZ06] for details on the Gleeble-tests.

From the measured data one can compute the evolution of the phase fraction and the strain caused by transformation induced plasticity e_{trip} (without assuming a model for TRIP!). In the case of a complete transformation one can then calculate the Greenwood-Johnson parameter κ and then obtains the evolution of the saturation function Φ, see again [WBDH08] for details. Their results encouraged Wolff et. al. to formulate a non-convex saturation function. The saturation functions we will deal with were already mentioned in Section 2.3.4, but for convenience we repeat:

(4.2.2a)	Tanaka	$\Phi(p) = p$
(4.2.2b)	Leblond	$\Phi(p) = p(1 - \ln(p))$
(4.2.2c)	Denis/Desalos	$\Phi(p) = p(2 - p)$
(4.2.2d)	Wolff	$\Phi(p) = \dfrac{1}{2}\left(1 + \sin\left(\pi p - \dfrac{\pi}{2}\right)\right)$

4.2.2 Verification of phase transformation models

In this subsection we will explain in detail how model parameters for phase transformation are obtained from experiments.

As this work is a continuation of [SFHW09, WBDH08], we use for the optimisation basically the programs of Isabel Hüßler but updated the material parameters to the SFB 570 material data set published in [ADF+08a, ADF+08b] and also improved the optimisation procedure.

As already explained above, the Gleeble-device can do tests in tension or compression and measures length change, diameter change, temperature and stress. Because the used specimen is a small and thin walled cylinder, we assume spatial homogeneity within the specimen.

From these measured quantities we can calculate a "measured" phase fraction, as it is described in [WB06], which later is used to optimise model parameters of the phase transformation models. There are two basic formulas for length and diameter change:

$$(4.2.3) \qquad \epsilon_l(t) = \frac{1}{E(\theta(t), p(t))} S(t) + \left(\sqrt[3]{\frac{\rho_0}{\rho(\theta(t), p(t))}} - 1 \right) + e_{trip}(t)$$

$$(4.2.4) \qquad \epsilon_d(t) = \frac{-\nu(\theta(t), p(t))}{E(\theta(t), p(t))} S(t) + \left(\sqrt[3]{\frac{\rho_0}{\rho(\theta(t), p(t))}} - 1 \right) - \frac{1}{2} e_{trip}(t)$$

where ϵ_l, ϵ_d are length and diameter change respectively, θ the temperature, p the phase fraction, E the Young modulus, ν the Poisson ratio, S the stress, ρ the density, ρ_0 the initial density and e_{trip} the inelastic strain caused by transformation induced plasticity.

Using the so-called bulk modulus, $K = \frac{E}{3(1-2\nu)}$, we can now introduce the volume strain:

$$(4.2.5) \qquad \epsilon_v(t) := \epsilon_l(t) + 2(t)\epsilon_d = \frac{1}{3K(\theta(t), p(t))} S(t) + 3 \left(\sqrt[3]{\frac{\rho_0}{\rho(\theta(t), p(t))}} - 1 \right)$$

As a next step we assume two mixture rules for the bulk modulus and the density, where we denote the austenite with p_1 and the forming martensite with p_2:

$$(4.2.6) \qquad \frac{1}{K(\theta, p)} = \frac{1}{K_1(\theta)} p_1 + \frac{1}{K_2(\theta)} p_2 \qquad \rho(\theta, p) = \rho_1(\theta) p_1 + \rho_2(\theta) p_2$$

Remembering that always holds $p_1 + p_2 = 1$, we obtain using (4.2.5) and (4.2.6) the following approximation for the volume strain at the discrete measured times t_i (we use $p(t_{i-1})$ in the mixture rule for the bulk modulus):

$$(4.2.7) \qquad \epsilon_v(t_i) \approx \frac{S(t_i)}{3} \left(\frac{1}{K_1(\theta(t_i))} + \left(\frac{1}{K_2(\theta(t_i))} - \frac{1}{K_1(\theta(t_i))} \right) p_2(t_{i-1}) \right) +$$
$$+ 3 \left(\sqrt[3]{\frac{\rho_0}{\rho_1(\theta(t_i)) + (\rho_2(\theta(t_i)) - \rho_1(\theta(t_i))) p_2(t_i)}} - 1 \right)$$

Solving this equation for $p_2(t_i)$ gives us an approximation for the martensitic phase fraction, which can be computed from the data measured by the Gleeble-device.

$$(4.2.8) \qquad p_2(t_i) \approx \frac{\rho_1(\theta(t_i))}{\rho_1(\theta(t_i)) - \rho_2(\theta(t_i))} \left(1 - \frac{\rho_0}{\rho_1(\theta(t_i))} \left[1 + \frac{\epsilon_v(t_i)}{3} + \right. \right.$$
$$\left. \left. - \frac{S(t_i)}{9} \left(\frac{1}{K_1(\theta(t_i))} + \left(\frac{1}{K_2(\theta(t_i))} - \frac{1}{K_1(\theta(t_i))} \right) p_2(t_{i-1}) \right) \right]^{-3} \right)$$

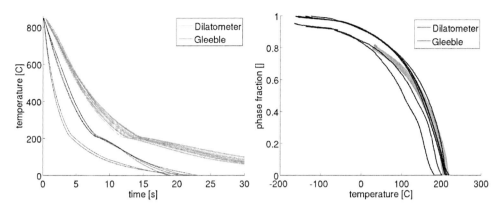

Figure 4.1: Temperature development and measured phase fraction for dilatometer and Gleeble test used for the optimisation of martensite models.

Equation (4.2.8) can be applied after the measured temperature has dropped under the martensite start temperature.

In the optimisation strategy the parameters of the different models are obtained as follows: For all experiments the L^2-norm of the errors between the "measured" phase fractions and the phase fractions predicted by the respective model are summed up and minimised (via optimisation of the model parameter).

In the material data set of the SFB 570, for the Koistinen-Marburger model there is already the model parameter available, obtained via a different optimisation approach. We use this parameter in all simulations, but did before an optimisation of the parameter to check, whether we would obtain similar results using a different optimisation approach.

The experimental data used for the optimisation are stress-free Gleeble tests (also used for the fitting of the Koistinen-Marburger model in [ADF+08b]) and dilatometer tests. In both types of experiments the used specimen are small, so that we can assume spatial homogeneity. Also, in both cases there is no external stress applied and equation (4.2.8) is used to calculate the "measured" phase fraction. In Figure 4.1 it can be seen, that the tests have different cooling rates, the dilatometer specimen are quenched more slowly than the Gleeble specimen. The resulting differences in the measured phase fraction are in accordance with the recently experimentally verified dependency of the martensitic transformation on the cooling rate, compare Section 4.1 for details.

The optimisation process for each martensite model and a set of n tests is conducted as follows. Via a variation of the corresponding model parameter, $mp \in \{\theta_{m0}, \mu, \beta\}$, the sum over all n tests of the L^2-error between the measured phase fraction, p_{data}, and the

dilatometer	K-M (opt)	K-M (SFB)	L-D	Yu
used parameter:	$\frac{1}{\theta_{m0}} = 0.0092$	$\frac{1}{\theta_{m0}} = 0.0107$	$\mu = 2.8496$	$\beta = 0.2145$
$\sum$ of L^2-error	0.4768	1.1392	1.1671	0.2918
$\varnothing$ of L^2-error	0.0207	0.0495	0.0507	0.0127
norm. SD	0.2915	0.1564	0.1414	0.4428

Gleeble	K-M (opt)	K-M (SFB)	L-D	Yu
used parameter:	$\frac{1}{\theta_{m0}} = 0.0116$	$\frac{1}{\theta_{m0}} = 0.0107$	$\mu = 10.434$	$\beta = 0.1601$
$\sum$ of L^2-error	0.1957	0.2113	0.2167	0.2281
$\varnothing$ of L^2-error	0.0326	0.0352	0.0361	0.0380
norm. SD	0.8153	0.4540	0.4016	0.4517

Table 4.2: Dilatometer and Gleeble tests: Results of the optimisation procedure for the Koistinen-Marburger, Leblond-Devaux and Yu model. Using equation (4.2.9) the sum and mean of the L^2-error as well as the normalised standard-deviation are given.

phase fraction predicted by the model, p_{cal}^{mp}, is minimised.

$$(4.2.9) \qquad \min_{mp} \sum \left(\frac{1}{T} \int_0^T \left(p_{data}(t) - p_{cal}^{mp}(t) \right)^2 dt \right)^{\frac{1}{2}}$$

For the optimisation the program Matlab® was used and the minimisation was done with the function `fminsearch`.

The results of the optimisation process for the two sets of experimental data are shown in Table 4.2. For all models the model parameter is specified together with the sum of the L^2-error over all experiments, the average L^2-error and its normalised standard-deviation.

Let us first discuss the results for the dilatometer tests. For the Koistinen-Marburger model a comparison between the optimised model parameter and the SFB value from [ADF+08b] was done. The optimisation provided a lower value than specified in [ADF+08b] and was able to halve the error. The optimisation of the model of Leblond-Devaux is quite unstable and yielded an approximation to the data slightly worse than the Koistinen-Marburger model. The model of Yu could be fitted very well to the dilatometer test and gave by far the smallest error of all three models.

Considering the Gleeble tests, the optimisation of the Koistinen-Marburger model results in a higher value than specified in [ADF+08b], here the error of the optimised case is only slightly lower. The error obtained by optimisation for the Leblond-Devaux model is slightly higher, than the error for the Koistinen-Marburger model. The model of

Yu gives an error a little bit higher than the Koistinen-Marburger and Leblond-Devaux model. For the Gleeble test the error obtained for all models are very similar.

In summary we see, that the model of Leblond-Devaux results in errors similar but higher than the Koistinen-Marburger model. For the dilatometer tests the model of Yu provides a considerably smaller error than the Koistinen-Marburger and Leblond-Devaux model while for the Gleeble tests its error is slightly higher. This result emphasises the need of including the isothermal martensitic transformation and the austenitic stabilisation into the modelling.

After the comparison in one-dimensional situations, we now come to a comparison with three-dimensional experiments

4.2.3   Validation: Comparison with experiments

In this subsection we want to use the above mentioned models for simulations of heat treatment of work-pieces and compare the simulated results with the measured ones. For the comparison of models for martensitic phase transformation and transformation induced plasticity we choose a setting where (nearly) no classical plasticity occurs, as this facilitates the interpretation of the results.

We deal with a conical ring which is symmetrically quenched in a gas nozzle field from $850°$C to $20°$C. The heat-transfer coefficients are $467\frac{W}{m^2K}$ outside, $438\frac{W}{m^2K}$ inside, $334\frac{W}{m^2K}$ at top and $217\frac{W}{m^2K}$ bottom. This quenching is strong enough to guarantee that only martensite is forming. Nevertheless it is also moderate enough to cause almost no classical plasticity (this is due to the symmetry of the problem, in additional simulations its effect was about 1%).

Experiments (compare [FLHZ09]) showed, that after the quenching the ring's outwall is not perpendicular but slightly inclined inside, see Figures 4.2 and 4.3. This effect can only be explained by transformation induced plasticity, since simulations without transformation induced plasticity result in a perpendicular outwall. The conical ring has a height of 26mm and an outer radius of 74mm. From the bottom to a height of 0.62mm the inner radius is 66.5mm while the inner radius at the top is 71.5mm. So in this setting we have a substantial influence of phase transition and transformation induced plasticity on the end deformation while we can neglect classical plasticity. These circumstances are well-suited for testing models for phase transformation and transformation induced plasticity.

Before we investigate the influence of the martensite models and of the saturation function on the inclination of the ring, we will do some numerical testing. We choose

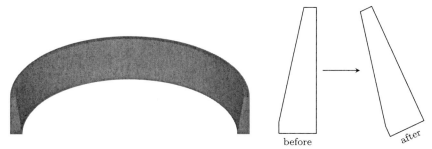

Figure 4.2: Cut through the conical ring

Figure 4.3:   Inclination after
the quenching

the martensite model of Koistinen-Marburger and the saturation function of Leblond and
compare the influence of three different meshes and of two different tolerances for time
step size control on the inclination of the outwall of the ring. The meshes we use are
structured meshes. They are build out of (deformed) cubes which are then divided into
six tetrahedra. A mesh can be characterised by the number of cubes in radial direction, n_r,
in circumferential direction, n_ϕ, and in height, n_h. The meshes can be seen in comparison
in Figure 4.4. Due to the symmetry of the problem, we use periodic boundary conditions,
as described in Section 3.6.1. In fact our geometry consists only of four lattice planes in
circumferential direction.

In Figure 4.5 the results of the comparison between the different meshes and tolerances
for time step size control are shown. In 4.5(a) we see the development of the spatial error
estimate of the deformation equation, compare Box 3.7.4, for the three different meshes
at tolerance 0.2. The coarsest mesh, mesh number one, has an error twice as high as
mesh number two, while the difference between the errors of meshes number two and
three is small. In 4.5(b) we see the time step size and the time error for mesh number
two and tolerances 0.2 and 0.15. The time error, $\eta_{\tau,n}$, is calculated using the equation
in Box 3.7.2 and the time step size control is implemented as described in Box 3.7.1,
with $\Gamma_\tau = 0.2, \alpha_1 = 1, \alpha_2 = 0.3$. For a tolerance of 0.2 the time error has to be smaller
than $\Gamma_\tau \alpha_1 tol = 0.04$ and the time step size is increased if the time error drops under
$\Gamma_\tau \alpha_2 tol = 0.012$. The time step size increases at the beginning of the simulation until the
phase transformation starts. Then it decreases drastically and again increases slowly till
the end of the simulation. Figures 4.5(c) and 4.5(d) depict the end deformation of the
outwall of the ring, the height over the change of diameter, and the angle of inclination for
the different meshes and tolerances of time step size control.It can be seen, that although
the differences in the inclination are small, mesh number one slightly overestimates the

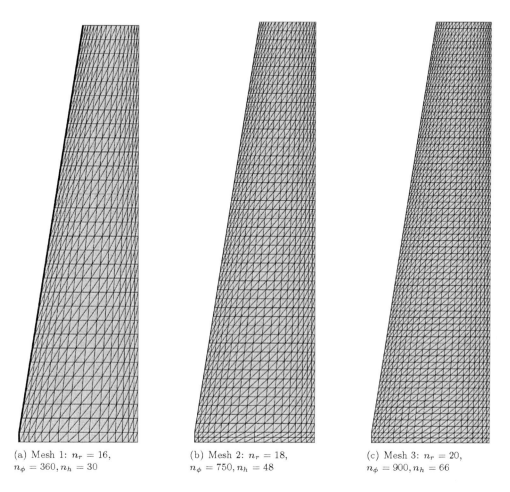

(a) Mesh 1: $n_r = 16$, $n_\phi = 360, n_h = 30$

(b) Mesh 2: $n_r = 18$, $n_\phi = 750, n_h = 48$

(c) Mesh 3: $n_r = 20$, $n_\phi = 900, n_h = 66$

Figure 4.4: Cross-section of the meshes used for numerical testing.

angle of inclination. Both tolerances for time step size control yield nearly equal results. In the following we will use for all simulations mesh number two with a time step size tolerance of 0.2.

After the numerical testing, we will now combine the three already mentioned marten- site models in (4.2.1) and the four saturation functions in (4.2.2). In Figures 4.6(a) to 4.6(d) we see the height over the change of diameter for the different models using the specified model parameters. Figure 4.6(e) depicts the inclination angle of the different simulations, while in Figure 4.6(f) the relative error between the measured inclination

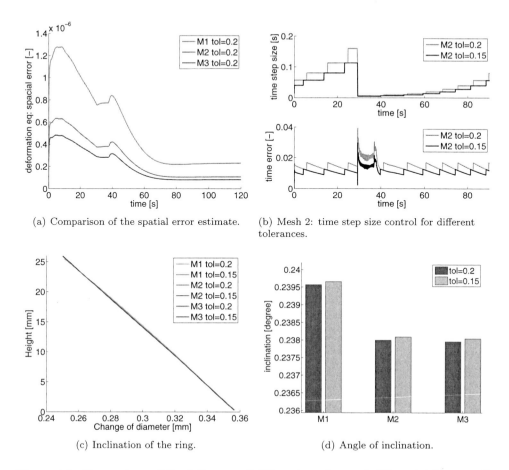

(a) Comparison of the spatial error estimate.

(b) Mesh 2: time step size control for different tolerances.

(c) Inclination of the ring.

(d) Angle of inclination.

Figure 4.5: Comparison of the influence of different meshes and different tolerances for time step size control on the inclination of the conical ring.

angle and the simulated one is shown. Beside the inclination of the outwall it is also interesting to compare the absolute change of diameter between simulations and experimental measurement. The absolute change of diameter depends directly on the fraction of rest-austenite at room temperature.

In X-ray measurements the fraction of rest-austenite at $20°C$ was found to be 13%. This was used as a side condition in the optimisation process for the model parameter of the Koistinen-Marburger model in [ADF+08b]. Therefore in all simulations using the Koistinen-Marburger model the simulated change of diameter is close to the measured one. The model of Leblond-Devaux also predicts the fraction of rest-austenite well, here

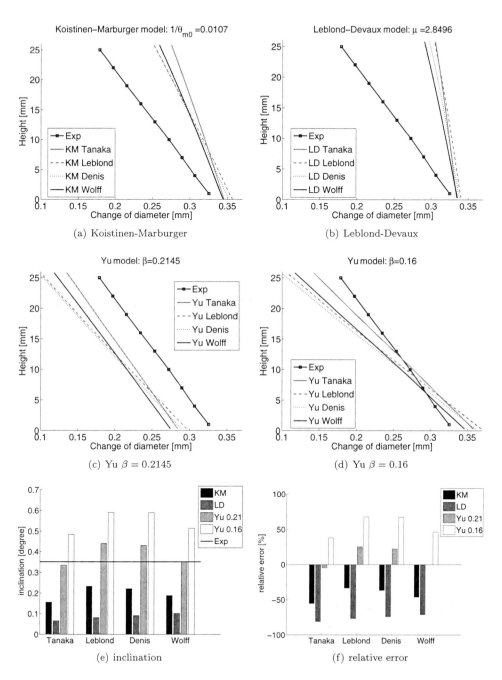

Figure 4.6: Comparison between simulated and measured data for the conical ring, considering the martensite models in (4.2.1) and the saturation functions in (4.2.2). Shown are the height over the change of diameter, the inclination angles, α, and the relative error between measured and simulated inclination angle, $\frac{\alpha - \alpha_{exp}}{\alpha_{exp}}$.

again the simulated change of diameter is close to the measured one. For the Yu model we compare simulations with model parameters $\beta = 0.2145$ and $\beta = 0.16$, the corresponding fractions of rest-austenite are 18% and 14%.

The Koistinen-Marburger model underestimates the measured inclination of the ring considerably, the relative errors between the simulated and measured inclination angle lie between -33% and -55%. The best result is obtained using the saturation function of Leblond.

The approximation of the inclination given by the Leblond-Devaux model is noticeably bad, the relative errors range between -71% and -81%. Here only the results for the simulations with $\mu = 2.8496$ are shown, the results with $\mu = 10.434$ are even less accurate.

For the Yu model the inclination for the simulations with $\beta = 0.16$, which was found by optimisation with Gleeble tests, are considerably to large. Their relative error in the inclination angle range between 37% and 67%. The simulations with $\beta = 0.2145$, which was found by optimisation with the more slowly quenched dilatometer tests, yields very good agreement with the experimental data. Here the relative error in the inclination angle lie between -0.1% and 25%. The best result is obtained using the saturation function of Wolff but also the saturation function of Tanaka yields very good results with -4%.

All in all it shows, that not only the different martensite models but even their model parameters have a very strong effect on the result of the simulation. The influence of the different saturation functions on the relative error in the inclination angle ranges between 10% and 30%.

Keeping in mind the newly found experimental results on martensitic transformation, see Section 4.1, we emphasise that the above results are only valid for this special setting of experiments and no generalisations are possible. The sensitivity of the problem to the martensitic transformation shows the need of including the new effects of martensitic transformation into the modelling. This will be subject of future work.

4.3   Inertia term

In this section we will investigate the influence of the inertia term in the deformation equation, which was neglected in the previous simulations, compare Section 2.5.

After deriving the discretised form of the non-stationary deformation equation, we will present an artificial testing situation with instantly released forces which cause oscillations. This is meant as a test whether the implementation works qualitatively correct. Then we will repeat the simulation of the conical ring from the previous section including the inertia term to test whether the result is influenced by it.

The discretisation of the stationary deformation equation is given in Box 3.3.2. The non-stationary deformation equation is a hyperbolic partial differential equation of second order. One possibility of discretisation for this type of equation is described in [QV97] and we proceed analogously. We begin with the spatial discretisation. In accordance with Section 3.3, we set

$$\boldsymbol{u}(t) = \sum_{i=1}^{3N} \boldsymbol{u}^i(t)\phi_i \in V_h^3, \qquad t \in (0,T)$$

Then the deformation equation discretised with respect to the space variables reads as follows:

$$\sum_{i=1}^{3N} \underbrace{\int_\Omega \rho\phi_i\phi_j dx}_{=:M_{ij}} \frac{\partial^2 \boldsymbol{u}^i}{\partial t^2}(t) + \sum_{i=1}^{3N} \underbrace{\int_\Omega 2\mu\epsilon(\phi_i):\epsilon(\phi_j) + \lambda\,\mathrm{div}(\phi_i)\,\mathrm{div}(\phi_j)dx}_{=:A_{ij}} \boldsymbol{u}^i(t) =$$

$$\underbrace{\int_{\partial\Omega} \boldsymbol{F}_{ext}\cdot\phi_j ds + \int_\Omega K\left(\frac{\rho_0-\rho}{\rho}\right)\boldsymbol{I}:\nabla\phi_j dx + \int_\Omega 2\mu(\epsilon^{up}+\epsilon^{cp}):\nabla\phi_j dx}_{=:F_j}$$

$$,j = 1,\dots,3N,\ t \in (0,T)$$

Using the more compact notation with mass matrix, $\boldsymbol{M}$, and stiffness matrix, $\boldsymbol{A}$, we can write the problem as an initial value problem.

$$\boldsymbol{M}\frac{\partial^2 \boldsymbol{u}}{\partial t^2}(t) + \boldsymbol{A}\boldsymbol{u}(t) = \boldsymbol{F}(t)$$

(4.3.1)
$$\boldsymbol{u}(0) = 0$$

$$\frac{\partial \boldsymbol{u}}{\partial t}(0) = 0$$

For time discretisation we consider as a start the following ordinary differential equation.

$$y''(t) = \psi(t, y(t), y'(t))$$

(4.3.2)
$$y(0) = y^0$$

$$y'(0) = y^1$$

Here $\psi : (0,T) \times \mathbb{R}^{3N} \times \mathbb{R}^{3N} \to \mathbb{R}^{3N}$ is a continuous function. The time discretisation of equations of type (4.3.2) can be done with the Newmark method. This method is formulated here for constant time step sizes, $t_n = t_{n-1} + \tau$. We denote by y_n and z_n the approximations of $y(t_n)$ and $y'(t_n)$ respectively. With $y_0 = y^0$ and $z_0 = y^1$ one can solve the following system of equations:

$$
(4.3.3) \qquad
\begin{aligned}
y_n &= y_{n-1} + \tau z_{n-1} + \tau^2 \left(\zeta \psi_n + \left(\frac{1}{2} - \zeta \right) \psi_{n-1} \right) \\
z_n &= z_{n-1} + \tau \left(\vartheta \psi_n + (1 - \vartheta) \psi_{n-1} \right)
\end{aligned}
$$

where $\psi_n := \psi(t_n, y_n, z_n)$ and $\zeta \geq 0$ and $\vartheta \geq 0$ are parameters, which determine the numerical properties of the Newmark method. If $\zeta = \vartheta = 0$, then the scheme (4.3.3) is explicit, otherwise it is implicit. In the case of $\zeta = \frac{1}{4}$ and $\vartheta = \frac{1}{2}$ the Newmark method is unconditionally stable.

When ψ does not depend on $y'(t)$, then it is possible to combine both equations of scheme (4.3.3) to one equation which depends only on y.

$$
(4.3.4) \quad y_{n+1} - 2y_n + y_{n-1} = \tau^2 \left(\zeta \psi_{n+1} + \left(\frac{1}{2} - 2\zeta + \vartheta \right) \psi_n + \left(\frac{1}{2} + \zeta - \vartheta \right) \psi_{n-1} \right)
$$

The above equation for time discretisation can now be applied to equation (4.3.1) and doing so, we obtain the discretised form of the non-stationary deformation equation.

4.3.1. Discretised non-stationary deformation equation:

$$
\left(\frac{1}{\tau^2} \boldsymbol{M} + \boldsymbol{A}\zeta \right) \boldsymbol{u}_n =
$$

$$
\left(\frac{2}{\tau^2} \boldsymbol{M} - \left(\frac{1}{2} - 2\zeta + \vartheta \right) \boldsymbol{A} \right) \boldsymbol{u}_{n-1} - \left(\frac{1}{\tau^2} \boldsymbol{M} - \left(\frac{1}{2} + \zeta - \vartheta \right) \boldsymbol{A} \right) \boldsymbol{u}_{n-2} +
$$

$$
+ \zeta \boldsymbol{F}_n + \left(\frac{1}{2} - 2\zeta + \vartheta \right) \boldsymbol{F}_{n-1} + \left(\frac{1}{2} + \zeta - \vartheta \right) \boldsymbol{F}_{n-2}
$$

Note that in the right hand side for each $\boldsymbol{F}_n$ the contribution from classical plasticity and transformation induced plasticity are taken from the previous time step t_{n-1}.

This non-stationary deformation equation is used in the simulations within this section.

4.3.1 Verification: Artificial testing situation

To begin with we will look at a simple situation: A long and thin cylinder ($l = 1m, d = 0.1m$) and homogeneous Dirichlet boundary conditions at both sides is considered. The

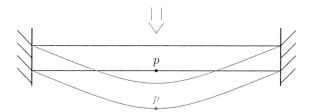

Figure 4.7: Artificial testing situation for the inertia term.

situation is isothermal, there is no phase transformation and the material behaves purely elastic. A volume force is applied to this cylinder which increases linearly, then stays constant for a short time and is then instantly released. See Figure 4.7 for a drawing of the setting.

The evolution of the volume force is shown in Figure 4.8(a). In Figure 4.8(b) we can see the displacement of a point p at the middle of the cylinder over the time. There are compared the results of two simulations using the stationary and the non-stationary deformation equation. In the non-stationary simulation the application of the volume force leads to a small oscillation in the displacement. This oscillation remains unchanged while the force grows and stays constant. When the volume force is released instantly, the resulting big oscillations superimpose with the small oscillations. Our model for the mechanical behaviour does neither include internal friction nor friction with the surround-

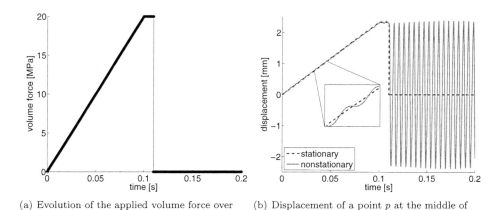

(a) Evolution of the applied volume force over time

(b) Displacement of a point p at the middle of the cylinder over time

Figure 4.8: Artificial testing situation for the inertia term.

ing. Therefore, the described behaviour is correct, and we can assume that the inertia term in our implementation works correctly. For a more realistic simulation one would have to include internal friction into the model. This would result in a damping of the oscillations. We do not consider internal friction, since the modelling would be difficult and there are no material parameters available. Moreover we will see in the next section, that in the settings we consider the non-stationary simulations do not differ considerably from the stationary simulations.

4.3.2   Validation: Application to a work piece experiment

After checking that our implementation of the inertia term works correctly we will now repeat the simulation of the conical ring from the previous section including the inertia term. We choose the simulation with the Koistinen-Marburger model and the Leblond saturation function.

In Figure 4.9 there is shown the evolution of the displacement in x-direction of both simulations at the point $p = (0.074, 0.0, 0.013)$. We can see that in the beginning of the quenching there is nearly no difference between the stationary and the non-stationary simulation. With the beginning of the phase transformation an oscillation starts. This could be caused by the transformation induced plasticity, whose values are taken from the previous time step in the discretisation. Over the whole time of the process the maximal deviation between the displacement of both simulations is about 0.033mm. For a realistic simulation the internal friction would have to be included into the model. This would result in a damping of the already small oscillations. Therefore we can conclude, that the inertia term can be neglected in the settings we are considering.

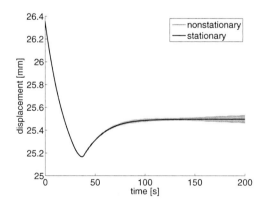

Figure 4.9: Application of the inertia term to the simulation if the quenching of a conical ring. Displacement over time at the point $p = (0.074, 0.0, 0.013)$

4.4 Classical plasticity

In this section we will deal with different models for isotropic and kinematic hardening of classical plasticity. Under consideration are the models of Ramberg-Osgood and
Prager and the model of Armstrong-Frederick which we already introduced in Section
2.4. Within the SFB 570 the model of Ramberg-Osgood and Prager is used. For the SFB
570 material data set, compare [ADF$^+$08a, ADF$^+$08b], tension tests were used to fit the
model parameters. We will comment on this process of parameter identification for the
Ramberg-Osgood and Prager model in the first subsection. Later it became possible to
conduct cyclic tension and compression tests and it showed that with the parameters fitted to the tension test the cyclic tension and compression tests were approximated poorly.
Therefore we will fit the Armstrong-Frederick model to those cyclic tests and will thus
try to improve the approximation. This process is described in Subsection 4.4.2. In the
following subsection we will compare both models in different situations.

4.4.1 Verification: Parameter identification for the Ramberg-Osgood and Prager model

In this subsection we will shortly explain the optimisation procedure for the model parameters of the hardening model of Ramberg-Osgood and Prager as it was done in the
SFB 570, compare [ADF$^+$08a, ADF$^+$08b].

Within the SFB 570 the commercial FE-Software Sysweld$^\text{®}$ is used for most of the
simulations involving classical plasticity. This software is relatively easy to use but has
strong restrictions concerning the models which can be used. There are different models for hardening available and it was chosen to use the model of Ramberg-Osgood for
isotropic hardening and the model of Prager for kinematic hardening. In the case of combined isotropic and kinematic hardening it is necessary to specify the fraction of isotropic
hardening. This is a major difference to the approach followed in this thesis.

For the Ramberg-Osgood model there were only the two model parameters determined,
$c(\theta), m(\theta)$. The parameter c_{cp} of the Prager model is internally calculated in Sysweld$^\text{®}$
from the data of the Ramberg-Osgood model via: for a fixed s_{cp}: $c_{cp}(\theta) = \frac{d}{dt}R(\theta, s_{cp})$.

The parameters of the Ramberg-Osgood model were determined from isothermal tension tests. These tests were conducted on supercooled austenite between 250°C and 850°C
and for martensite they were conducted on annealed martensite between 20°C and 200°C.

Contrary to the approach in this work, the classical Ramberg-Osgood model includes
no yield stress. In this case even small applied stresses cause a plastic deformation. For
one dimensional tension tests, $(s^{cp} = \epsilon^{cp})$, the relation between stress and strain is as

follows:

(4.4.1) $S = c(\theta)(\epsilon^{cp})^{m(\theta)}$

(4.4.2) $\epsilon = \epsilon^{el} + \epsilon^{cp} = \dfrac{S}{E} + \left(\dfrac{S}{c(\theta)}\right)^{\frac{1}{m(\theta)}}$

At first one has to determine the Young modulus, E, via fitting a line to the beginning of the stress strain curve, whose slope is E. Then one can use equation (4.4.2) for the optimisation of $c(\theta)$ and $m(\theta)$. This was done separately for each isothermal test. A linear dependence on the temperature was assumed and an interpolation over the temperature through the data of the isothermal tests was done. (For the parameter c for austenite there are also two non-linear dependencies from the temperature specified, but we will not consider this here.) The results are:

austenite: $c(\theta) = -1.034\theta + 960.7$ [MPa] $m = 0.1092$ [-]

martensite: $c(\theta) = -6.76\theta + 5724.1$ [MPa] $m = 8.449 \cdot 10^{-5}\theta + 0.1278$ [-]

As in Sysweld® it is convenient to specify a yield stress, the following formula was used to define the offset yield stress:

(4.4.3) $R_0(\theta) := c(\theta)(0.00005)^{m(\theta)}$

The main problem in considering tension tests only is, that nothing can be said about the fraction of isotropic and kinematic hardening. The complicated interaction of isotropic and kinematic hardening can only be studied in cyclic tension and compression tests as we will see in the next section.

4.4.2   Verification: Parameter identification for the Armstrong-Frederick model

In this subsection we will explain in detail how the model parameters for the Armstrong-Frederick model are obtained via optimisation. For this purpose we use isothermal cyclic tension and compression tests of supercooled austenite which were conducted on the Gleeble-device, compare [DIL+09, ŞDL+10]. At the moment there are no tension and compression tests available for martensite. To do these tests for martensite would be very costly. The specimen would need to be heat treated before the Gleeble test, to ensure that it consists of martensite completely. Then the specimen would be re-heated

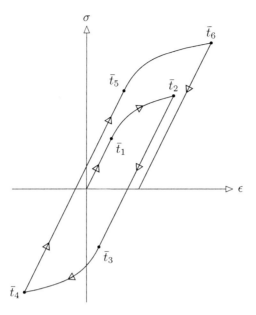

Figure 4.10: Drawing of a cyclic tension and compression test with marked times of beginning and end of plastic flow.

for the Gleeble test. Until now in all simulations done for the SFB 570, classical plasticity occurred only in the austenitic phase, as it is softer than martensite. Also the yield stress increases with decreasing temperature, the yield stress of austenite at martensite start temperature is about eight times higher then in the beginning of the simulation at 850°C. So from this point of view a restriction of the knowledge of the hardening behaviour to the austenitic phase will not cause problems in the near future.

The isothermal cyclic tension and compression tests of austenite are strain controlled. There is a sketch of the stress-strain diagram of the experiments in Figure 4.10. The measured quantities are time, temperature, length change and diameter change. The model of Armstrong-Frederick has four model parameters, $\hat{c}, \hat{b}(\theta), \hat{\gamma}, \hat{\beta}(\theta)$, we allow two of them to depend linearly on the temperature. At first we will explain our optimisation strategy for these parameters. The basic idea is to take the measured strain and to calculate the stress predicted by the Armstrong-Frederick model for a specific parameter set. The resulting stress-strain curve is then compared to the measured one. The optimisation procedure tries to find a set of parameters which minimise the difference between the measured and the calculated curve. The optimisation process is applied to each isothermal test separately. Depending of the quality of the results, it is then possible to obtain

a temperature dependent set of parameters via interpolation. In Box 4.4.1 we summarise
our optimisation strategy.

4.4.1. Optimisation strategy for the parameters of the Armstrong-Frederick model:

- *parameter optimisation of $\hat{c}, \hat{b}, \hat{\gamma}, \hat{\beta}$ for each isothermal tension and compression test separately*

    - *given: measured data $(t_{exp}, \theta_{exp}, S_{exp}, \epsilon_{exp})$*
    - *choose initial values $\hat{c}_0, \hat{b}_0, \hat{\gamma}_0, \hat{\beta}_0$*
    - *optimisation process:*
      find a set of parameters $\hat{c}, \hat{b}, \hat{\gamma}, \hat{\beta}$, which minimise:

      (4.4.4) $$||S_{exp} - S_{AF}(\hat{c}, \hat{b}, \hat{\gamma}, \hat{\beta})||_{L^2}$$

- *formulation of a temperature dependent set of parameters $\hat{c}, \hat{b}(\theta), \hat{\gamma}, \hat{\beta}(\theta)$*

Now we will explain how the stress predicted by the Armstrong-Frederick model can
be calculated. The length change, ϵ, is measured by the Gleeble-device during the time
of the experiment. After we scaled the length change so that it starts with zero, we take
one value after another to calculate all needed parameters: ϵ^{el} the elastic strain, ϵ^{cp} the
plastic strain, s^{cp} the accumulated plastic strain, X the back-stress, R the increase of
the yield surface and S the stress. It is worth noting, that the information necessary
for this purpose consists of the Young modulus, the yield stress, the measured strain
and temperature. Although the test should be isothermal there are fluctuations in the
temperature, that is why we will consider the Young modulus and the yield stress to be
temperature dependent. For this method of optimisation it is not necessary to specify
the areas of plastic yielding in advance, as this is figured out be the method itself.

In the beginning of the experiment all plastic quantities, $\epsilon^{cp}, s^{cp}, X, R$, are set to zero.
Then we use in every time step t_i the yield function to decide, whether we are in an elastic
or a plastic time step.

(4.4.5) $$F := |(\epsilon_i - \epsilon_{i-1}^{cp})E(\theta_i) - X_{i-1}| - R_0(\theta_i) - R_{i-1}$$

If F is smaller than zero, then we are in an elastic time step. Otherwise we are in a plastic time step. Here we need to find expressions for X_i, R_i and ϵ_i^{cp} depending on S_i, so that in the yield condition, (4.4.6), the variable S_i is the only unknown.

$$(4.4.6) \qquad 0 = |(\epsilon_i - \epsilon_i^{cp}(S_i))E(\theta_i) - X_i(S_i)| - R_0(\theta_i) - R_i(S_i)$$

These expressions are different for the three plastic branches, compare Figure 4.10, between $(\bar{t}_1, \bar{t}_2)$, $(\bar{t}_3, \bar{t}_4)$ and $(\bar{t}_5, \bar{t}_6)$. We will deduce these expressions now in detail and then specify the algorithm used for the computation of S_i in Box 4.4.2. At first we calculate expressions for the accumulated plastic strain on the three plastic branches. Here it is important to remember that the plastic quantities do not change during the elastic parts, therefore it holds $\epsilon^{cp}(\bar{t}_2) = \epsilon^{cp}(\bar{t}_3)$ and $\epsilon^{cp}(\bar{t}_4) = \epsilon^{cp}(\bar{t}_5)$.

$$(4.4.7a) \qquad \bar{t}_1 \le t \le \bar{t}_2 : \; s^{cp}(t) = \int_0^t |\dot{\epsilon}^{cp}(\tau)| d\tau = \epsilon^{cp}(t)$$

$$(4.4.7b) \qquad \bar{t}_3 \le t \le \bar{t}_4 : \; s^{cp}(t) = \int_{\bar{t}_1}^{\bar{t}_2} \dot{\epsilon}^{cp}(\tau) d\tau - \int_{\bar{t}_3}^t \dot{\epsilon}^{cp}(\tau) d\tau$$

$$= \epsilon^{cp}(\bar{t}_2) - (\epsilon^{cp}(t) - \epsilon^{cp}(\bar{t}_2))$$

$$= 2\epsilon^{cp}(\bar{t}_2) - \epsilon^{cp}(t)$$

$$(4.4.7c) \qquad \bar{t}_5 \le t \le \bar{t}_6 : \; s^{cp}(t) = 2\epsilon^{cp}(\bar{t}_2) - \epsilon^{cp}(\bar{t}_4) + (\epsilon^{cp}(t) - \epsilon^{cp}(\bar{t}_4))$$

$$= 2\epsilon^{cp}(\bar{t}_2) - 2\epsilon^{cp}(\bar{t}_4) + \epsilon^{cp}(t)$$

Without phase transformation and for constant temperature the hardening is given by the following two equations.

$$(4.4.8) \qquad \dot{X}(t) = \hat{c}\dot{\epsilon}^{cp}(t) - \hat{b}X(t)\dot{s}^{cp}(t)$$

$$(4.4.9) \qquad \dot{R}(t) = \hat{\gamma}\dot{s}^{cp}(t) - \hat{\beta}\dot{s}^{cp}(t)R(t)$$

It can be easily verified that the solutions of these two equations are as follows:

$$(4.4.10) \qquad X(t) = \hat{c}e^{-\hat{b}s^{cp}(t)} \int_0^t \dot{\epsilon}^{cp}(\tau)e^{\hat{b}s^{cp}(\tau)} d\tau$$

$$(4.4.11) \qquad R(t) = \frac{\hat{\gamma}}{\hat{\beta}}\left(1 - e^{-\hat{\beta}s^{cp}(t)}\right)$$

Using equations (4.4.7) we can now derive formulas for the back stress in the three plastic branches.

$\bar{t}_1 \le t \le \bar{t}_2:$

$$s^{cp}(t) = \epsilon^{cp}(t)$$

$$X(t) = \hat{c}e^{-\hat{b}\epsilon^{cp}(t)} \int_0^t \dot{\epsilon}^{cp}(\tau)e^{\hat{b}\epsilon^{cp}(\tau)}d\tau$$

$$= \hat{c}e^{-\hat{b}\epsilon^{cp}(t)} \left[\frac{1}{\hat{b}}e^{\hat{b}\epsilon^{cp}(\tau)}\right]_0^t$$

(4.4.12a)
$$= \frac{\hat{c}}{\hat{b}}\left(1 - e^{-\hat{b}\epsilon^{cp}(t)}\right)$$

$\bar{t}_3 \le t \le \bar{t}_4:$

$$s^{cp}(t) = 2\epsilon^{cp}(\bar{t}_2) - \epsilon^{cp}(t)$$

$$X(t) = \hat{c}e^{s^{cp}(t)}\left(\int_0^{\bar{t}_2} \dot{\epsilon}^{cp}(\tau)e^{\hat{b}\epsilon^{cp}(\tau)}d\tau + \int_{\bar{t}_3}^{\bar{t}_4} \dot{\epsilon}^{cp}(\tau)e^{\hat{b}(2\epsilon^{cp}(\bar{t}_2)-\epsilon^{cp}(\tau))}d\tau\right)$$

$$= \frac{\hat{c}}{\hat{b}}e^{-\hat{b}(2\epsilon^{cp}(\bar{t}_2)-\epsilon^{cp}(t))}\left(e^{\hat{b}\epsilon^{cp}(\bar{t}_2)} - 1 - e^{-\hat{b}(2\epsilon^{cp}(\bar{t}_2)-\epsilon^{cp}(t))} + e^{\hat{b}\epsilon^{cp}(\bar{t}_2)}\right)$$

(4.4.12b)
$$= \frac{\hat{c}}{\hat{b}}\left(-1 - e^{-\hat{b}(2\epsilon^{cp}(\bar{t}_2)-\epsilon^{cp}(t))} + 2e^{-\hat{b}(\epsilon^{cp}(\bar{t}_2)-\epsilon^{cp}(t))}\right)$$

$\bar{t}_5 \le t \le \bar{t}_6:$

$$s^{cp}(t) = 2\epsilon^{cp}(\bar{t}_2) - 2\epsilon^{cp}(\bar{t}_4) + \epsilon^{cp}(t)$$

$$X(t) = \hat{c}e^{s^{cp}(t)}\left(\frac{2}{\hat{b}}e^{\hat{b}\epsilon^{cp}(\bar{t}_2)} - \frac{1}{\hat{b}} - \frac{2}{\hat{b}}e^{-\hat{b}(2\epsilon^{cp}(\bar{t}_2)-\epsilon^{cp}(\bar{t}_4))} + \right.$$

$$\left. + \frac{1}{\hat{b}}e^{-\hat{b}(2\epsilon^{cp}(\bar{t}_2)-\epsilon^{cp}(\bar{t}_4)+\epsilon^{cp}(t))}\right)$$

(4.4.12c)
$$= \frac{\hat{c}}{\hat{b}}\left(1 + 2e^{-\hat{b}(\epsilon^{cp}(\bar{t}_2)-2\epsilon^{cp}(\bar{t}_4)+\epsilon^{cp}(t))} - e^{-\hat{b}(2\epsilon^{cp}(\bar{t}_2)-2\epsilon^{cp}(\bar{t}_4)+\epsilon^{cp}(t))} + \right.$$

$$\left. - 2e^{-\hat{b}(\epsilon^{cp}(t)-\epsilon^{cp}(\bar{t}_4))}\right)$$

In Box 4.4.2, we describe how the stress predicted by the Armstrong-Frederick model is calculated from a given strain. The main idea is to put equations (4.4.7),(4.4.11) and (4.4.12) into equation (4.4.6) and to replace $\epsilon^{cp}(t)$ by $\epsilon^{cp}(t) = \epsilon(t) - \epsilon^{el}(t) = \epsilon(t) - \frac{S(t)}{E}$. In this way one obtains an equation for the determination of the stress in the plastic branches.

4.4.2. Algorithm for the calculation of the stress corresponding to a given strain using the Armstrong-Frederick model:

- *given: measured data:* (t, θ, ϵ)

- *Initialise all quantities*

$$\epsilon_1^{el} = 0, \epsilon_1^{cp} = 0, s_1^{cp} = 0, X_1 = 0, R_1 = 0, S_1 = 0$$

- *For all time steps* t_i

  - **If** $|(\epsilon_i - \epsilon_{i-1}^{cp})E - X_{i-1}| - R_0(\theta) - R_{i-1} < 0$ *elastic time step:*

    $$\epsilon_i^{el} = \epsilon_i - \epsilon_{i-1}^{cp}$$
    $$S_i = (\epsilon_i^{el})E_i$$
    $$\epsilon_i^{cp} = \epsilon_{i-1}^{cp} \quad X_i = X_{i-1} \quad R_i = R_{i-1} \quad s_i^{cp} = s_{i-1}^{cp}$$

  - **Else** *plastic time step:*

    * *calculate* S_i *by solving numerically:*
    $\bar{t}_1 \leq t_i \leq \bar{t}_2$: *first plastic branch*

    $$0 = \left| S_i - \frac{\hat{c}}{\hat{\gamma}} \left(1 - e^{-\hat{b}\left(\epsilon_i - \frac{S_i}{E_i}\right)} \right) \right| - R_{0,i} - \frac{\hat{\gamma}}{\hat{\beta}} \left(1 - e^{-\hat{\beta}\left(\epsilon_i - \frac{S_i}{E_i}\right)} \right)$$

    $\bar{t}_3 \leq t_i \leq \bar{t}_4$: *second plastic branch*

    $$0 = \left| S_i - \frac{\hat{c}}{\hat{\gamma}} \left(-1 - e^{-\hat{b}(2\epsilon^{cp}(\bar{t}_2) - \left(\epsilon_i - \frac{S_i}{E_i}\right))} \right. \right.$$
    $$\left. \left. + 2e^{-\hat{b}(\epsilon^{cp}(\bar{t}_2) - \left(\epsilon_i - \frac{S_i}{E_i}\right))} \right) \right| +$$
    $$- R_{0,i} - \frac{\hat{\gamma}}{\hat{\beta}} \left(1 - e^{-\hat{\beta}(2\epsilon^{cp}(\bar{t}_2) - \left(\epsilon_i - \frac{S_i}{E_i}\right))} \right)$$

$\bar{t}_5 \leq t_i \leq \bar{t}_6$: *third plastic branch*

$$0 = \left| S_i - \frac{\hat{c}}{\hat{b}}\left(1 + 2e^{-\hat{b}(\epsilon^{cp}(\bar{t}_2) - 2\epsilon^{cp}(\bar{t}_4) + \left(\epsilon_i - \frac{S_i}{E_i}\right))} + \right.\right.$$

$$\left. - e^{-\hat{b}(2\epsilon^{cp}(\bar{t}_2) - \epsilon^{cp}(\bar{t}_4) + \left(\epsilon_i - \frac{S_i}{E_i}\right))} + \right.$$

$$\left.\left. - 2e^{-\hat{b}(\left(\epsilon_i - \frac{S_i}{E_i}\right) - \epsilon^{cp}(\bar{t}_4))}\right)\right| +$$

$$- R_{0,i} - \frac{\hat{\gamma}}{\hat{\beta}}\left(1 - e^{-\hat{\beta}(2\epsilon^{cp}(\bar{t}_2) - 2\epsilon^{cp}(\bar{t}_4) + \left(\epsilon_i - \frac{S_i}{E_i}\right))}\right)$$

∗ *update all remaining quantities*

$$\epsilon_i^{el} = \frac{S_i}{E_i}$$

$$\epsilon_i^{cp} = \epsilon_i - \epsilon_i^{el}$$

$$s_i^{cp} = \begin{cases} \epsilon^{cp}(t_i) & \bar{t}_1 \leq t_i \leq \bar{t}_2 \\ 2\epsilon^{cp}(\bar{t}_2) - \epsilon^{cp}(t_i) & \bar{t}_3 \leq t_i \leq \bar{t}_4 \\ 2\epsilon^{cp}(\bar{t}_2) - 2\epsilon^{cp}(\bar{t}_4) + \epsilon^{cp}(t_i) & \bar{t}_5 \leq t_i \leq \bar{t}_6 : \end{cases}$$

$$X_i = \begin{cases} \frac{\hat{c}}{\hat{b}}\left(1 - e^{-\hat{b}\epsilon^{cp}(t_i)}\right) & \bar{t}_1 \leq t_i \leq \bar{t}_2 \\ \frac{\hat{c}}{\hat{b}}\left(-1 - e^{-\hat{b}(2\epsilon^{cp}(\bar{t}_2) - \epsilon^{cp}(t_i))} + \right. & \\ \left. +2e^{-\hat{b}(\epsilon^{cp}(\bar{t}_2) - \epsilon^{cp}(t_i))}\right) & \bar{t}_3 \leq t_i \leq \bar{t}_4 \\ \frac{\hat{c}}{\hat{b}}\left(1 + 2e^{-\hat{b}(\epsilon^{cp}(\bar{t}_2) - 2\epsilon^{cp}(\bar{t}_4) + \epsilon^{cp}(t_i))} + \right. & \\ \left. -e^{-\hat{b}(2\epsilon^{cp}(\bar{t}_2) - 2\epsilon^{cp}(\bar{t}_4) + \epsilon^{cp}(t_i))} + \right. & \\ \left. -2e^{-\hat{b}(\epsilon^{cp}(t_i) - \epsilon^{cp}(\bar{t}_4))}\right) & \bar{t}_5 \leq t_i \leq \bar{t}_6 : \end{cases}$$

$$R_i = \frac{\hat{\gamma}}{\hat{\beta}}\left(1 - e^{-\hat{\beta}s^{cp}(t_i)}\right)$$

We use the above algorithm to calculate for one experiment the corresponding stress predicted by the Armstrong-Frederick model for a given set of model parameters.

The experimental data available for the fitting of the parameters are six isothermal tension and compression tests at the following temperatures: 300°C, 400°C, 650°C, 700°C, 750°C and 850°C. The stress strain curves of these experiments are plotted in Figure 4.11.

Due to capacity reasons there are only these six experiments available. This is a problem for the parameter fitting process because for a reliable result more experiments

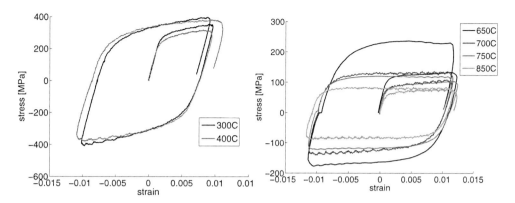

Figure 4.11: Isothermal tension and compression tests used for the fitting of the Armstrong-Frederick model.

at different temperatures and repeated tests are necessary. The Gleeble machine has general problems with reproducibility of measurements. At tension tests with a different steel sort deviations in the measured stress of up to 10MPa occurred. Additionally there are some other problems with some of the experiments.

The experiments at high temperatures, especially the test at 850°C, could involve creep effects. This can not be cleared without experiments with different strain rate, which are not conducted, until now.

The stress strain curve of the experiment at 700°C is almost identical with the stress strain curve of the experiment at 750°C. This could be linked to the problems with reproducibility, already mentioned.

In the third plastic branch of the test at 650°C the temperature rose at 14°C.

In the third plastic branch of the test at 300°C there could be phase transformation present.

Despite these problems we will use these experimental data for parameter fitting, as there is no other data available. We start with a fitting of parameters to each single experiment as described in Box 4.4.1. For the optimisation we use the Matlab® routine `lsqcurvefit`. Numerical optimisation procedures search a *local* minimum of the considered function starting from the given initial values. Therefore we choose for each of the four parameters three different values, compare (4.4.13), and compare the optimisation results corresponding to the $3^4 = 81$ initial values.

$$\hat{c} = [5000, 10000, 20000]\, \text{MPa}$$
$$\hat{b} = [500, 1250, 2000]$$
(4.4.13)
$$\hat{\gamma} = [2500, 5000, 10000]\, \text{MPa}$$
$$\hat{\beta} = [10, 100, 200]$$

The optimisation results for the different initial values for each experiment can be found in Figure 4.12. For each tension and compression test the error given by (4.4.4) (sorted increasingly) and the values of the four parameters $\hat{c}, \hat{b}, \hat{\gamma}, \hat{\beta}$ are plotted. It can be seen that for the experiment at $300°$C the results for different initial values vary considerably. For all other data sets there are some initial values which lead to a minimal error and to similar or equal parameters.

In Figure 4.13 we see two results of the fitting process. For the temperatures $400°$C and $650°$C there are plotted the measured stress strain curve, the stress strain curve of the Armstrong-Frederick model, the back stress, $\boldsymbol{X}^{cp}$, over the plastic strain and the increase of the yield surface, R, over the accumulated plastic strain. It can be seen, that the Armstrong-Frederick model is able to approximate the measured data quite well. It is worth noting that the development of the back stress is highly non-linear, in contrast to the Ramberg-Osgood and Prager model, where the back stress is assumed to be linear.

Figure 4.14 depicts the optimisation results of the four model parameters for all six isothermal experiments. As one can see, the results scatter considerably. Therefore a simple averaging or interpolation will provide no good results in obtaining a temperature dependent set of parameters. If there were more experiments available, most probably it would be easier to see trends for the parameters. As this is not the case, we have to try out some values for the parameters within the scatter band of the results to find a temperature dependent set of parameters which give a good approximation of the Armstrong-Frederick model to the experimental data. This temperature dependent set of parameters is plotted as dashed lines in Figure 4.14, the values are as follows:

(4.4.14) $\hat{c} = 18400\, [\text{MPa}] \quad \hat{b} = 0.79\theta + 39.9 \quad \hat{\gamma} = 1840\, [\text{MPa}] \quad \hat{\beta} = 0.13\theta - 34.7$

For simplicity only $\hat{b}$ and $\hat{\beta}$ are considered to be temperature dependent. It is also possible to choose $\hat{c}$ and $\hat{\gamma}$ temperature dependent. This would result in two additional terms in the equations for the back stress and the increase of the yield surface. As the

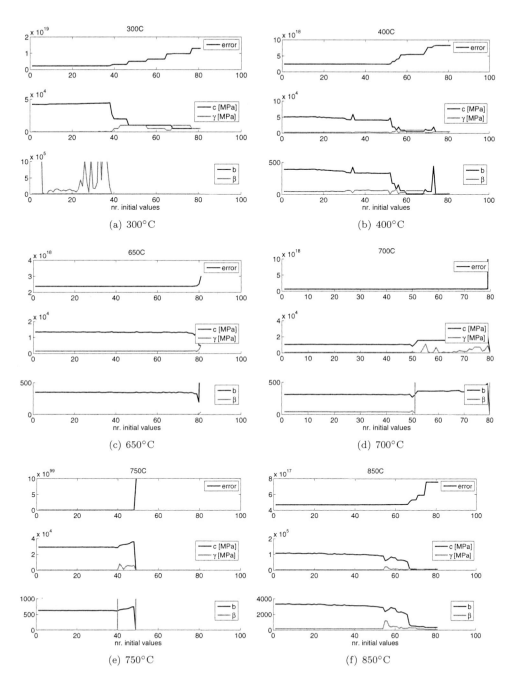

Figure 4.12: Optimisation results for the different initial values for each tension and compression test.

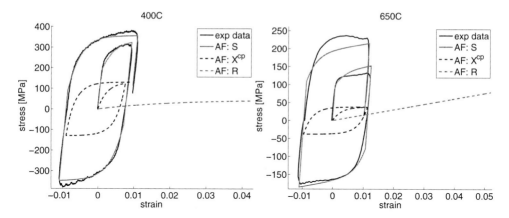

Figure 4.13: Comparison between experimental data and the fitted Armstrong-Frederick model. Plotted are the stress strain curve of the Armstrong-Frederick model, the back stress, $\boldsymbol{X}^{cp}$, over the plastic strain and the increase of the yield surface, R over the accumulated plastic strain.

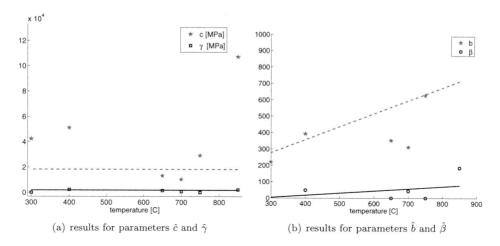

Figure 4.14: Results of the optimisation process for the different temperatures (symbols). Temperature dependent set of parameters as in (4.4.14) (dashed).

Armstrong-Frederick model is more sensitive on $\hat{b}$ and $\hat{\beta}$ and as long as we do not have more experiments which could indicate a temperature dependence, we will consider $\hat{c}$ and $\hat{\gamma}$ to be constant.

In Figure 4.15 there is shown a comparison of the Ramberg-Osgood model (isotropic hardening only), the Ramberg-Osgood and Prager model and the Armstrong-Frederick model using the isothermal tension and compression tests. For the Armstrong-Frederick

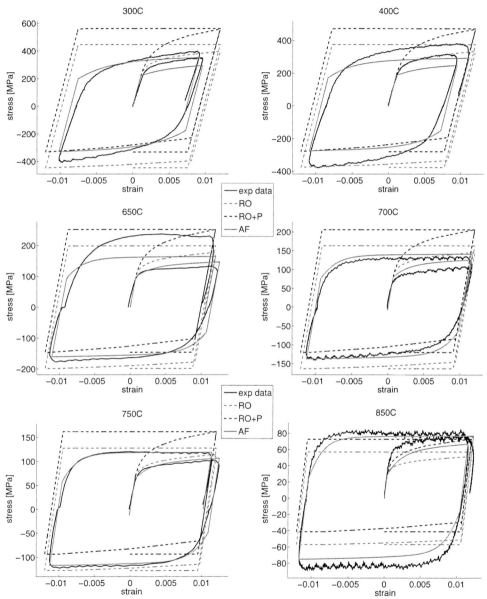

Figure 4.15: Comparison of the Ramberg-Osgood model (isotropic hardening only), Ramberg-Osgood and Prager model and the Armstrong-Frederick model using isothermal cyclic tension and compression tests.

model the temperature dependent parameter set (4.4.14) is used together with the algorithm described in Box 4.4.2. For the Ramberg-Osgood model and the Ramberg-Osgood and Prager model there are used three-dimensional FE-simulations where a similar strain path as in the experiments is applied. Please note, that the third plastic branch of the test at 650°C was not taken into account for the comparison of models, as here temperature deviations occurred.

We also consider purely isotropic hardening with the Ramberg-Osgood model, to see whether this model approximates well the hardening behaviour in the first plastic branch. This is the case for those tests, where the first plastic branch is similar to the tension tests, which were used to determine the parameters of the Ramberg-Osgood model in [ADF+08a]. Naturally the approximation of those tension and compression tests is bad, which differ considerably from the tension tests used for the determination of the parameters of the Ramberg-Osgood model. It is clear, that a purely isotropic model gives a bad approximation of the second and third plastic branch, because the experimental data shows clearly the presence of kinematic hardening.

The Ramberg-Osgood and Prager model shows a bad approximation of the experimental results. Especially in the first plastic branch the stress is much to high. The reason for the bad approximation lies most probably in the way of parameter fitting in [ADF+08a]. Within the SFB 570 the FE-software Sysweld® is used. In the case of combined isotropic and kinetic hardening one has to specify the fraction of isotropic hardening. This is an essential difference to our approach, so that for a usage of the Ramberg-Osgood and Prager model in our setting a new fitting of parameters would be necessary. Simulations of the cyclic tension and compressions tests done within the SFB 570 with Sysweld® also show a bad approximation of the experimental data, especially of the second and third plastic branch. Therefore we do no extra parameter fitting for the Ramberg-Osgood and Prager model, but concentrate on the Armstrong-Frederick model.

It can be seen, that the Armstrong-Frederick model is able to approximate well the experimental data. Especially the points of re-yielding and the second and third plastic branches are approximated much better than by the other models. Considering the shortage of experimental data and the problems some of the test are harbouring, these results are encouraging. It seems probably that with a sufficient set of reliable experiments, the Armstrong-Frederick model could be used to improve the description of the hardening behaviour of 100Cr6.

4.4.3   Validation: Comparison of the two plasticity models

In this subsection we will compare qualitatively the two plasticity models of Ramberg-Osgood and Prager and of Armstrong-Frederick. In the previous subsection we already compared the two models on isothermal cyclic tension and compression tests. For a further comparison it would be ideal to have experimental data with in situ measurements of tests where classical plasticity occurs during quenching. Unfortunately such data is not available, so that we have to restrict our comparison to situations where differences between the models cause different simulation results, without being able to compare the simulation results to experimental data.

At first we will look at situations where the coupling between the back stresses of classical plasticity and transformation induced plasticity plays and important role. After dealing with small specimen, which are pre-hardened before quenching, we will consider conical rings, as described in Section 4.2.3, which are quenched so quickly, that classical plasticity occurs.

As a next step we will simulate more complex situations like an asymmetrically quenched cylindrical ring. For this experiment there are measurements of the deformation of the ring before and after the heat treatment. But there are several reasons why the simulation results can not be compared directly to the measurements, most important are the already in Section 4.2 mentioned problems in modelling the martensitic phase transformation. So in this setting as well we have to restrict ourselves to a qualitative comparison.

Back stress coupling in pre-hardening tests

In the literature there are experimental results on the coupling between the back stresses of classical plasticity and transformation induced plasticity. In [Ahr03, TP02, TPG06] there are conducted tests with small specimens which were plastically pre-hardened in tension or compression before the phase transformation started. Although during the phase transformation, there was no stress applied, a considerable TRIP strain occurred. There are experiments where a pre-hardening in tension leads to a negative TRIP strain and a pre-hardening in compression leads to a positive TRIP strain, but there are also experiments where a pre-hardening in tension leads to a positive TRIP strain and vice versa. In [Ahr03] an isothermal bainitic transformation shows a positive TRIP strain after pre-hardening in tension. In [TP02, TPG06] there is a different steel sort considered and here it is found that for the bainitic transformation a pre-hardening in tension leads to a negative TRIP strain while for martensitic transformation a pre-hardening in tension

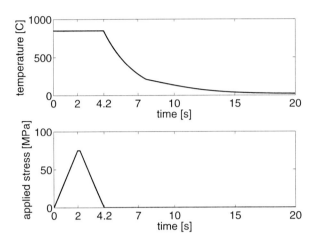

Figure 4.16: Setting of the pre-hardening simulations: While the temperature remains constantly 850°C the specimen are pre-hardened with different applied stresses (positive or negative). After unloading the specimen are quenched without applied stress.

leads to a positive TRIP strain. Moreover in [MBTS07] a numerical modelling of the martensitic transformation on the micro-scale is done. In this setting the direction of the TRIP strain after pre-hardening in tension depends on the orientation of the forming martensite plates. However experimental results for this dependency are not included in this paper.

In our approach both hardening models can show a positive or negative TRIP strain after pre-hardening in tension. This can be realised via a variation of the sign of the coupling parameter c_{int}. A positive c_{int} leads to a negative TRIP strain for pre-hardening in tension and a negative c_{int} leads to a positive TRIP strain for pre-hardening in tension. In the following we will simulate pre-hardening tests using small hollow probes ($r_{in} = 3$mm, $r_{out} = 4$mm, $h = 6$mm). Because these specimen are small and the quenching using a heat transfer coefficient of $900 \frac{W}{m^2 K}$ in- and outside and $100 \frac{W}{m^2 K}$ on top and bottom is quick, we can assume, that the temperature inside the specimen is homogeneous. This means that in cases without pre-hardening and applied stress during phase transformation, there will be no transformation induced plasticity. The setting of the simulations is as follows: At first the specimen are isothermally pre-hardened at 850°C and then quenched to 20°C, undergoing a martensitic transformation. In Figure 4.16 the evolution of the temperature and the applied stress can be seen.

At the end of the simulation the occurring TRIP strain is calculated in the way as it is done in [TPG06]. Thereby only the length strain of our three-dimensional simulation is considered. We use the additive composition of the strain tensor:

$$(4.4.15) \qquad \epsilon^{up} = \epsilon - \epsilon^{cp} - \epsilon^{thm} - \epsilon^{el}$$

where ϵ^{up} is the TRIP strain, ϵ is the total strain, ϵ^{cp} is the plastic strain, ϵ^{thm} is the thermal strain and ϵ^{el} is the elastic strain. As we calculate the TRIP strain at the end of the test, where no external stress is applied, the elastic strain is zero. The thermal strain is determined via a simulation without pre-hardening, where only the quenching and the phase transformation take place. The plastic strain is determined before the quenching of the specimen starts: $\epsilon^{cp} = \frac{\Delta l}{l} - \frac{S}{E}$. In this way the TRIP strain was computed for all simulations.

For a simulation of the interaction of the back stresses of classical plasticity and transformation induced plasticity, we need the parameters c_{up} for the TRIP back stress and the coupling parameter c_{int}. As there are no experiments available for determining the parameter c_{up} for the martensitic phase transformation on 100Cr6 steel, we use a parameter similar to the one obtained for pearlite on 100Cr6 steel in [WBD+06]: $c_{up} = 2000$MPa.

For both hardening models the maximal coupling parameter c_{int} was computed, such that for the Ramberg-Osgood and Prager model $c_{int}^2 \leq c_{cp} c_{up}$ holds and for the Armstrong-Frederick model $c_{int}^2 \leq \hat{c} c_{up}$. As already remarked in Section 2.4 these conditions arise from the thermodynamical way of modelling classical plasticity and assure that the free inelastic energy is convex for frozen temperature and phases. Because the parameter c_{cp} in the Ramberg-Osgood and Prager model is temperature dependent and becomes small for high temperatures and growing plastic accumulation, the allowed maximal coupling parameter is relatively small: $|c_{int}| \leq 500$MPa. As the parameter $\hat{c}$ of the Armstrong-Fredrick model is constant a considerably stronger coupling is allowed: $|c_{int}| \leq 6000$MPa.

Simulations are conducted for both hardening models using the allowed maximal positive and negative coupling parameter and for different positive and negative applied stresses in pre-hardening. The results can be seen in Figure 4.17. It shows that with a negative c_{int} the TRIP strain has the same direction as the stress applied in pre-hardening while for a positive c_{int} it has the opposite direction. Thus our model presented is capable of reproducing the experimentally investigated interactions of back stresses.

Moreover it can be seen in Figure 4.17 that the Ramberg-Osgood and Prager model shows a linear dependence between plastic strain and TRIP strain. By contrast, the Armstrong-Frederick model shows a strong non-linear dependence, which is qualitatively

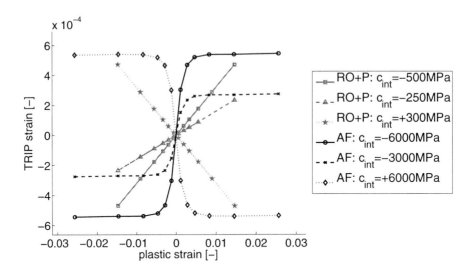

Figure 4.17: Results of the pre-hardening simulations with Ramberg-Osgood and Prager model or Armstrong-Frederick model, using $c_{up} = 2000$MPa and c_{int} as specified. Each group of simulation was pre-hardened applying the following stresses: 75, 70, 65, 60, 55, 45, -45, -55, -60, -65, -70, -75 MPa. The TRIP strain is calculated using equation (4.4.15).

in good agreement with the experimental results in [Ahr03].

Remark. It is worth noting, that the Leblond model for classical plasticity and transformation induced plasticity also includes a coupling between both effects. However this model is only capable of reproducing the effect, that pre-hardening in tension causes a negative TRIP strain, compare [MBTS07, TPG06, WBH08].

Back stress coupling in three-dimensional settings

As a next step we want to compare the two hardening models and the influence of the coupling between classical plasticity and transformation induced plasticity in a less artificial setting with three-dimensional stress states. Therefore we use again the geometry of the conical ring, already described in Section 4.2.3, but this time we choose the heat transfer coefficients high enough to ensure that classical plasticity occurs. In all our simulations the rings are quenched from 850°C to 20°C. For our considerations we set the heat transfer coefficient inside the ring equal to the one outside, $\delta_{in} = \delta_{out}$, and the heat transfer coefficient on the top equals the one on the bottom, $\delta_{top} = \delta_{bot}$. Depending on the ratio of these heat transfer coefficients, $\frac{\delta_{in}}{\delta_{top}}$, the outwall of the ring is more or less inclined at the end of the simulation. We will compare four different ratios of the heat transfer coefficients:

Ratio:	1:1	3:2	3:1	10:1
$\delta_{in} \left[\frac{W}{m^2 K}\right]$	2000	2000	2000	2000
$\delta_{top} \left[\frac{W}{m^2 K}\right]$	2000	1400	700	200

At first we use the above heat transfer coefficients in simulations with either the Ramberg-Osgood and Prager or the Armstrong-Fredrick model, while the transformation induced plasticity has no back stress and thus no coupling between classical and transformation induced plasticity is considered, $c_{up} = c_{int} = 0$. The resulting inclination of the outwall of the ring and the inclination angle can be seen in Figure 4.18. For both hardening models we can observe the effect, that for the heat transfer coefficient ratio 1:1 the inclination is very small. The inclination increases with an increasing ratio and reaches its maximum at the ratio 10:1. In the simulation with ratio 1:1 there is considerably more classical plasticity present, than in the simulation ratio 10:1. Therefore we can conclude, that in this setting classical plasticity counteract transformation induced plasticity. For the calculation of the angle of inclination, the interpolating line of the simulation results was used. In Figure 4.18(b) the inclination angle of the simulations using the Armstrong-Frederick and the Ramberg-Osgood model are shown. For the simulations using the same heat transfer coefficient ratio, there is also specified the relative deviation between the angle of the simulations using Armstrong-Frederick and the Ramberg-Osgood model, $\Delta := \frac{\alpha_{AF} - \alpha_{ROP}}{\alpha_{AF}}$. It shows, that for all four simulations using the Ramberg-Osgood and Prager model the resulting inclination is smaller than in the simulations using the Armstrong-Frederick model. The relative deviation between both hardening models reduces with decreasing plasticity, and is almost negligible at the ratio of 10:1.

In the next step we compare the influence of the back stress coupling on the resulting inclination. As already mentioned in the preceding subsection, there are no experiments available for determining the parameter c_{up} for the martensitic phase transformation on 100Cr6 steel. Therefore we use a parameter similar to the one obtained for pearlite on 100Cr6 steel in [WBD+06]: $c_{up} = 2000$MPa. As explained before the allowed coupling parameters are for the Ramberg-Osgood and Prager model $|c_{int}| \leq 500$MPa and for the Armstrong-Fredrick model $|c_{int}| \leq 6000$MPa. In Figure 4.19 it is shown for the Armstrong-Frederick model and the Ramberg-Osgood and Prager model the relative deviation between the angles of inclination of the simulation with and without back stress coupling between classical plasticity and transformation induced plasticity, e.g. the blue bar stands for the relative deviation between the angles $\frac{\alpha_{(c_{up}=c_{int}=0)} - \alpha_{(c_{up}=2e9Pa,c_{int}=0)}}{\alpha_{(c_{up}=c_{int}=0)}}$.

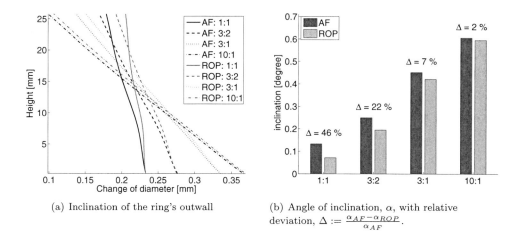

(a) Inclination of the ring's outwall

(b) Angle of inclination, α, with relative deviation, $\Delta := \frac{\alpha_{AF} - \alpha_{ROP}}{\alpha_{AF}}$.

Figure 4.18: Deformation of the outwall of the conical ring after simulations using different ratios of heat transfer coefficients, $\frac{\delta_{in}}{\delta_{top}}$, and using the hardening models of Armstrong-Frederick or Ramberg-Osgood and Prager.

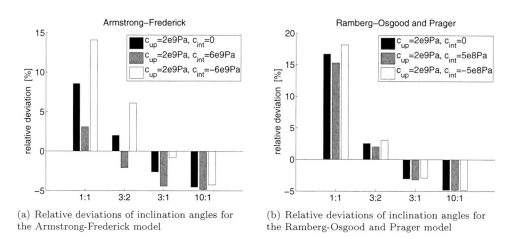

(a) Relative deviations of inclination angles for the Armstrong-Frederick model

(b) Relative deviations of inclination angles for the Ramberg-Osgood and Prager model

Figure 4.19: Comparison between the Armstrong-Frederick model and the Ramberg-Osgood and Prager model. Shown are the relative deviations between the angles in Figure 4.19 and the corresponding angles for simulations with back stress coupling, e.g. the blue bar stands for the relative deviation between the angles $\frac{\alpha_{(c_{up}=c_{int}=0)} - \alpha_{(c_{up}=2e9Pa, c_{int}=0)}}{\alpha_{(c_{up}=c_{int}=0)}}$.

For both hardening models it can be seen, that for heat transfer coefficient ratios, with much classical plasticity, the inclination increases when the transformation induced plasticity has its own back stress, $c_{up} = 2000\text{MPa}$. This effect decreases with decreasing classical plasticity and and the angle of inclination can even become smaller than in the case with, $c_{up} = 0$. A positive coupling parameter, $c_{up} = 2000\text{MPa}$ and $c_{int} > 0$, reduces the inclination compared to the case $c_{up} = 2000\text{MPa}$ and $c_{int} = 0$. A negative coupling parameter has the opposite effect. Although the absolute value of the allowed coupling parameter is for the Armstrong-Frederick model much higher than for the Ramberg-Osgood and Prager model, the influence of the coupling on the inclination angle is similar for both hardening models.

Finally we can conclude from the above comparison, that there are considerable differences in the simulation results, for the two hardening models of Armstrong-Frederick and of Ramberg-Osgood and Prager. Moreover we can state a qualitative influence of the back stress coupling between classical plasticity and transformation induced plasticity. To determine this influence quantitatively the material parameter c_{up} had to be determined from experimental data.

Simulation of asymmetrically quenched rings

As a last step we want to compare both hardening models in a setting, where experimental data is available. We consider a cylindrical ring, which has an outer radius of 72.5mm, an inner radius of 66.5mm and a height of 26mm. This ring is quenched asymmetrically in a gas nozzle field from 850°C to 20°C. The gas nozzle field consists of twelve nozzle holders inside and outside of the ring, compare Figure 4.20. Each nozzle holder has three nozzles in vertical direction. From inside the ring is slowly quenched, from outside there are only six nozzle holders active, which cause a quick quenching in certain areas of the ring, see Figure 4.21. After the quenching of the ring the outer radius is measured in circumferential direction at different heights. A measurement at the height of 13mm can be seen in Figure 4.22. Where the ring is quenched from outside, the radius is smaller than in the area, where the gas nozzles are turned off.

For the simulation we use symmetric boundary conditions. Considering the quenching situation it is sufficient to restrict the simulated geometry to a quarter in circumference and to a half in height. This setting is shown in the dark gray part in Figure 4.21. The heat transfer coefficients for the simulation was computed inside the SFB 570 in sub-project B4 with the commercial FE-program ANSYS® FLUENT®. The spatial variation of the heat transfer coefficients can be seen in Figure 4.23.

Figure 4.20:   Picture of the gas nozzle field

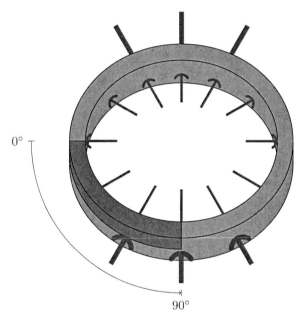

Figure 4.21:   Schematic representation of the asymmetrically quenched ring. The dark gray part shows the reduced geometry for the simulation using symmetric boundary conditions.

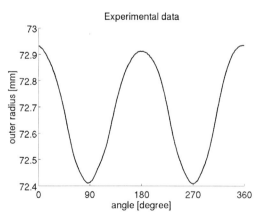

Figure 4.22:   Measured outer radius at height 13mm.

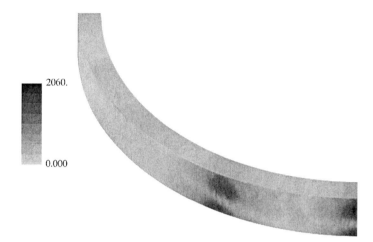

Figure 4.23: Heat transfer coefficients in $\left[\frac{W}{m^2K}\right]$ for the asymmetrically quenched ring.

Before we investigate the influence of the different hardening models on the ring's deformation, we will compare the results for different meshes and tolerances for time step size control. This is done to ensure, that the spatial resolution of the mesh and the time step sizes is fine enough. The meshes we use are structured and built out of deformed cubes which are then divided in six tetrahedra. A mesh can be characterised by the number of cubes in radial direction, n_r, in circumferential direction, n_ϕ, and in height, n_h. For this setting we use meshes which are uniform in height and circumference but in radial direction the elements become finer towards the outwalls. For the comparison of time step sizes we vary the overall tolerance tol and the parameter c_2 which belongs to the contribution of classical plasticity in the error indicator for time step size control, compare Box 3.7.2. All comparisons are made for simulations including transformation induced plasticity and classical plasticity with the hardening model of Ramberg-Osgood and Prager. The influence of the different meshes can be seen in Figure 4.24(a), where the outer radius at height 13mm is shown. The influence of different time step sizes is negligible, compare Figures 4.24(b) and 4.24(c). This is surprising as the semi-implicit scheme, which we developed for classical plasticity, is said to be less robust against big time steps than the fully implicit scheme. One possible explanation could be, that in this setting the deformation of the ring is mostly caused by transformation induced plasticity (which we will see later in the model comparison). This result is most probably not transferable to other situations with a stronger influence of classical plasticity.

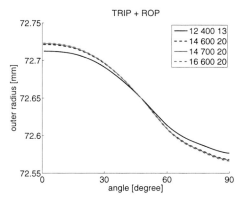

(a) Comparison of different meshes, the numbers in the legend specify n_r, n_ϕ, n_h. tol=0.55, c_2=1000.

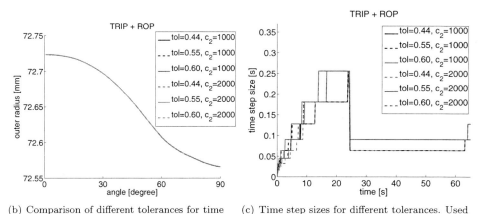

(b) Comparison of different tolerances for time step size control. Used mesh: $n_r = 16, n_\phi = 600, n_h = 20$

(c) Time step sizes for different tolerances. Used mesh: $n_r = 16, n_\phi = 600, n_h = 20$

Figure 4.24: Comparison of the use of different meshes and time step sizes.

For all following simulations we will use the mesh $n_r = 16, n_\phi = 600, n_h = 20$, tolerance 0.55 and $c_2 = 1000$. In Figure 4.26 it is shown the deformed ring in comparison to the undeformed mesh.

Now we will compare simulations with and without classical plasticity, with different hardening models and with and without back stress coupling. In Figure 4.25(a) we see the results of simulations with different hardening models but without back stress coupling. It can be seen that the radius change of the ring is mostly an effect of transformation induced

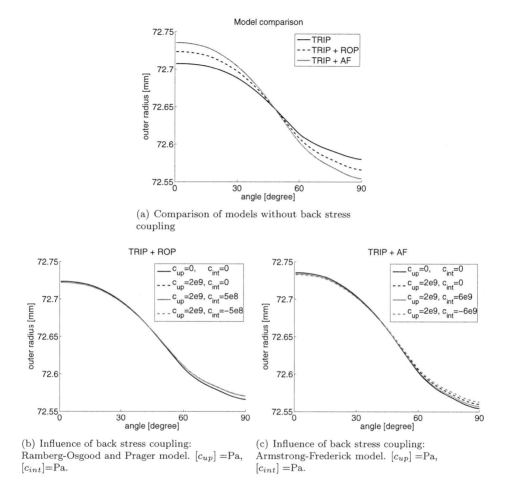

(a) Comparison of models without back stress coupling

(b) Influence of back stress coupling: Ramberg-Osgood and Prager model. $[c_{up}]$ =Pa, $[c_{int}]$=Pa.

(c) Influence of back stress coupling: Armstrong-Frederick model. $[c_{up}]$ =Pa, $[c_{int}]$ =Pa.

Figure 4.25: Model comparison: Used mesh: $n_r = 16, n_\phi = 600, n_h = 20$, tol=0.55, c_2=1000.

plasticity. Taking classical plasticity into account increases the radius change, here the Armstrong-Frederick model shows a bigger effect than the Ramberg-Osgood and Prager model. In Figures 4.25(b) and 4.25(c) there is shown a comparison with and without back stress coupling for the Ramberg-Osgood and Prager model and for the model of Armstrong-Frederick. For the Ramberg-Osgood and Prager model the allowed value for the coupling parameter is with $|c_{int}| \leq 500$MPa relatively small, and the influence of the back stress coupling on the result is negligible. For the Armstrong-Frederick model

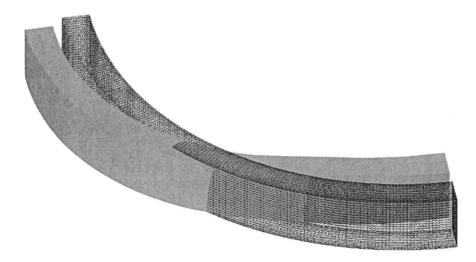

Figure 4.26: Deformation at the end of the quenching (increased by factor 50) and the undeformed (scaled) mesh.

the allowed value for the coupling parameter is higher, $|c_{int}| \leq 6000MPa$. Nevertheless the influence of the back stress coupling on the result is small. In the setting of the asymmetrically quenched ring the back stress coupling seems to be negligible.

In comparison to the measured data in Figure 4.22 the simulated change in radius is considerably too small. The amplitude of the measured radius change is about 0.5mm while the simulated radius change has an amplitude of about 0.17mm which is only a third of the measured value. The reason for this difference is probably related to the problems with the modelling of the martensitic phase transformation, as already mentioned in Section 4.1. This result emphasises the need for better modelling of the martensitic phase transformation.

CHAPTER

5

Outlook

Throughout this thesis we introduced a mathematical model for steel quenching, developed a numerical scheme and showed in simulations that the model is in accordance with reality. Naturally there remain open problems which we want to address in this chapter together with some remarks on future work.

It has already been mentioned in Section 4.1, that for a good agreement of the simulation result with experimental measurements, it is essential to model the phase transformation well. Therefore the martensitic isothermal transformation and the austenitic stabilisation have to be included into the modelling.

In the comparison of the two hardening models for classical plasticity in Section 4.4.3 it can be seen, that there is a considerable difference between both models. Especially for classical plasticity additional experiments would be needed for an further improvement of the simulations.

- For the model of Armstrong-Frederick a better experimental basis for the determination of the model parameters would be needed. This concerns also the austenitic phase but especially the martensite, as here are no experiments available at all.

125

- For a determination of the quantitative extent of the coupling between the back-stresses of classical and transformation induced plasticity, there would be needed experimental data for the determination of c_{up}, as they are presented in [WBD+06] for pearlite.

- The next step would be the design of Gleeble experiments where both classical and transformation induced plasticity occur and the coupling parameter c_{int} can be determined.

- Experiments of this kind are planed in a project of the German Research Foundation (DFG): "Mehr-Mechanismen-Modelle: Theorie und ihre Anwendung auf einige Phänomene im Materialverhalten von Stahl" (BO 1144/4-1). This project is a spin-off of the SFB 570. Here multi-mechanism models are applied to the case of small deformations. The main characteristic of these models is the additive decomposition of the inelastic strain (e.g. plastic or visco-plastic) into two (or multi) parts (sometimes called "mechanisms"). We refer to [CS95, TC10] for discussion and further references. The interaction between classical plasticity and transformation induced plasticity dealt with above is a concrete application of a two-mechanism approach (cf. [WBH08]).

Further steps for an improvement of the simulations will consist in including the heating. During this process there occur creep and transformation induced plasticity, so that the specimen is already deformed when the quenching starts.

The implementation could be enhanced by the development of a fully implicit algorithm for classical plasticity. Then the time steps could be chosen larger, which would reduce the computational effort.

Appendices

APPENDIX

A

Discretisation of the
Armstrong-Frederick model

This section deals with the numerical realisation of the third sub step of the overall solution scheme in Box 3.2.1, when the Armstrong-Frederick model is used for classical plasticity. The continuous equations of the Armstrong-Frederick model can be found in Box 2.4.3.

In our overall solution procedure we already solved the heat, phases and deformation equations for the new time step. Now we use the newly calculated displacement to decide, whether this time step is elastic or plastic. This can be achieved by computing the trial stress $\boldsymbol{\sigma}^t$ (equation (3.5.1)) and the trial yield function, see equation (3.5.2) and the explanations below.

Again the crucial point is the derivation of a scalar equation for the numerical calculation of the plastic multiplier. The yield condition will be used for this and therefore formulations of the norm of the effective stress and the increase of the yield surface are important.

We start with the easier part of gaining an expression for the increase of the yield function (for the single phases) R^i, where the only unknown is the plastic multiplier.

$$(A.1) \qquad R_n^i = R_{n-1}^i + \hat{\gamma}_i \tau_n \sqrt{\frac{2}{3}} \gamma_n - \hat{\beta}_i \tau_n R_n^i \sqrt{\frac{2}{3}} \gamma_n$$

$$(A.2) \qquad R_n^i = \frac{1}{1 + \hat{\beta}_i \tau_n \sqrt{\frac{2}{3}} \gamma_n} \left(R_{n-1}^i + \hat{\gamma}_i \tau_n \sqrt{\frac{2}{3}} \gamma_n \right)$$

To deal with the effective stress is more complicated. We begin with a repetition of the formula for the corrected stress.

$$\sigma_n^{c*} = 2\mu(\epsilon_n - \epsilon_n^{cp} - \epsilon_n^{up})$$

$$= 2\mu(\epsilon_n - \epsilon_{n-1}^{cp} - \epsilon_{n-1}^{up} - \tau_n \dot{\epsilon}_n^{cp} - \tau_n \dot{\epsilon}_n^{up})$$

$$(A.3) \qquad = \sigma_n^{t*} - 2\mu\tau_n\gamma_n \frac{\boldsymbol{\xi}_n^{cp}}{||\boldsymbol{\xi}_n^{cp}||} - 3\mu\tau_n\kappa\boldsymbol{\xi}_n^{up} \sum_i^N \dot{\Phi}_i(p_i)\dot{p}_i$$

Then we look at both back-stresses, using again the trick $\boldsymbol{X}_n = \boldsymbol{X}_{n-1} + \tau_n\dot{\boldsymbol{X}}$, see Box 2.4.3 for the continuous equations.

$$\boldsymbol{X}_n^{cp} = \boldsymbol{X}_{n-1}^{cp} + \frac{2}{3}\tau_n\hat{c}\gamma_n \frac{\boldsymbol{\xi}_n^{cp}}{||\boldsymbol{\xi}_n^{cp}||} - \sqrt{\frac{2}{3}}\gamma_n\hat{b}\tau_n\boldsymbol{X}_n^{cp} + c_{int}\tau_n\kappa\boldsymbol{\xi}_n^{up} \sum_i^N \dot{\Phi}_i(p_i)\dot{p}_i$$

$$(A.4) \quad \boldsymbol{X}_n^{cp} = \underbrace{\frac{1}{1 + \sqrt{\frac{2}{3}}\gamma_n\hat{b}\tau_n}}_{=:d_x} \left(\boldsymbol{X}_{n-1}^{cp} + \frac{2}{3}\tau_n\hat{c}\gamma_n \frac{\boldsymbol{\xi}_n^{cp}}{||\boldsymbol{\xi}_n^{cp}||} + c_{int}\tau_n\kappa\boldsymbol{\xi}_n^{up} \sum_i^N \dot{\Phi}_i(p_i)\dot{p}_i \right)$$

$$(A.5) \quad \boldsymbol{X}_n^{up} = \boldsymbol{X}_{n-1}^{up} + \frac{2}{3}\tau_n c_{int}\gamma_n \frac{\boldsymbol{\xi}_n^{cp}}{||\boldsymbol{\xi}_n^{cp}||} - \sqrt{\frac{2}{3}}\gamma_n c_{int}\frac{\hat{b}}{\hat{c}}\tau_n\boldsymbol{X}_n^{cp} + \tau_n c_{up}\kappa\boldsymbol{\xi}_n^{up} \sum_i^N \dot{\Phi}_i(p_i)\dot{p}_i$$

For convenience we remember the definition of the trial effective stresses $\boldsymbol{\xi}_n^{cp,t} = \sigma_n^{t*} - \boldsymbol{X}_{n-1}^{cp}$ and $\boldsymbol{\xi}_n^{up,t} = \sigma_n^{t*} - \boldsymbol{X}_{n-1}^{up}$ before we use equations (A.3), (A.4) and (A.5) to gain expressions for the effective stresses.

$$(A.6) \qquad \boldsymbol{\xi}_n^{cp} = \sigma_n^{t*} - d_x\boldsymbol{X}_{n-1}^{cp} - (2\mu + \frac{2}{3}d_x\hat{c})\tau_n\gamma_n \frac{\boldsymbol{\xi}_n^{cp}}{||\boldsymbol{\xi}_n^{cp}||} +$$

$$- (3\mu + d_x c_{int})\tau_n\kappa \left(\sum_i^N \dot{\Phi}_i(p_i)\dot{p}_i \right) \boldsymbol{\xi}_n^{up}$$

$$\boldsymbol{\xi}_n^{up} = \underbrace{\boldsymbol{\xi}_n^{up,t} + \sqrt{\frac{2}{3}}\gamma_n c_{int}\frac{\hat{b}}{\hat{c}}d_x\tau_n\boldsymbol{X}_{n-1}^{cp}}_{=:\boldsymbol{F}_1} +$$

$$- \underbrace{\left(2\mu + \frac{2}{3}c_{int} - \sqrt{\frac{2}{3}}\gamma_n c_{int}\hat{b}\tau_n\frac{2}{3}d_x\right)}_{=:f_2}\tau_n\gamma_n\frac{\boldsymbol{\xi}_n^{cp}}{||\boldsymbol{\xi}_n^{cp}||} +$$

$$- \underbrace{\left(3\mu + c_{up} - \sqrt{\frac{2}{3}}\gamma_n c_{int}^2\frac{\hat{b}}{\hat{c}}\tau_n d_x\right)}_{=:f_3}\tau_n\kappa\left(\sum_i^N \dot{\Phi}_i(p_i)\dot{p}_i\right)\boldsymbol{\xi}_n^{up}$$

(A.7) $$\quad \boldsymbol{\xi}_n^{up} = \frac{1}{1 + f_3\tau_n\kappa\left(\sum_i^N \dot{\Phi}_i(p_i)\dot{p}_i\right)}\left(\boldsymbol{F}_1 - f_2\tau_n\gamma_n\frac{\boldsymbol{\xi}_n^{cp}}{||\boldsymbol{\xi}_n^{cp}||}\right)$$

When we now put equation (A.7) into equation (A.6), we define $d := \frac{(3\mu + c_{int}d_x)\tau_n\kappa\left(\sum_i^N \dot{\Phi}_i(p_i)\dot{p}_i\right)}{1 + f_3\tau_n\kappa\left(\sum_i^N \dot{\Phi}_i(p_i)\dot{p}_i\right)}$ as abbreviation.

(A.8) $$\quad \boldsymbol{\xi}_n^{cp}\left(1 + (2\mu + \frac{2}{3}d_x\hat{c} - df_2)\tau_n\gamma_n\frac{1}{||\boldsymbol{\xi}_n^{cp}||}\right) = \boldsymbol{\sigma}_n^{t*} - d_x\boldsymbol{X}_{n-1}^{cp} - d\boldsymbol{F}_1$$

Next we apply the norm to the above equation.

$$||\boldsymbol{\xi}_n^{cp}||\left|1 + (2\mu + \frac{2}{3}d_x\hat{c} - df_2)\tau_n\gamma_n\frac{1}{||\boldsymbol{\xi}_n^{cp}||}\right| = ||\boldsymbol{\sigma}_n^{t*} - d_x\boldsymbol{X}_{n-1}^{cp} - d\boldsymbol{F}_1||$$

(A.9) $$\quad \Leftrightarrow \left|||\boldsymbol{\xi}_n^{cp}|| + \underbrace{(2\mu + \frac{2}{3}d_x\hat{c} - df_2)\tau_n\gamma_n}_{=:g_2}\right| = ||\underbrace{\boldsymbol{\sigma}_n^{t*} - d_x\boldsymbol{X}_{n-1}^{cp} - d\boldsymbol{F}_1}_{=:\boldsymbol{G}_1}||$$

Finally we can combine equation (A.9) with the yield condition:

(A.10) $$\quad F = 0 = ||\boldsymbol{\xi}_n^{cp}|| - \sqrt{\frac{2}{3}}(R_0 + R_n)$$

and obtain (together with (A.2)) a scalar equation in which the plastic multiplier is the only unknown and can be computed numerically.

(A.11) $$\quad \left|\sqrt{\frac{2}{3}}(R_0 + R_n(\gamma_n)) + g_2(\gamma_n)\tau_n\gamma_n\right| = ||\boldsymbol{G}_1(\gamma_n)||$$

After finding a possibility of calculating the plastic multiplier, we can formulate the algorithm for plasticity with isotropic and kinematic hardening using the Armstrong-Frederick model.

A.1.1. Algorithm for plasticity with the Armstrong-Frederick model:

- *Calculate trial stress*

$$\boldsymbol{\sigma}_n^{t*} = 2\mu\left(\boldsymbol{\epsilon}(\boldsymbol{u}_n) - \boldsymbol{\epsilon}_{n-1}^{cp} - \boldsymbol{\epsilon}_{n-1}^{up}\right)$$

$$\boldsymbol{\xi}_n^{cp,t} = \boldsymbol{\sigma}_n^{t*} - \boldsymbol{X}_{n-1}^{cp}$$

$$\boldsymbol{\xi}_n^{up,t} = \boldsymbol{\sigma}_n^{t*} - \boldsymbol{X}_{n-1}^{up}$$

- *Evaluate yield function, using trial effective stress and R_{n-1}*

$$F_n^t = ||\boldsymbol{\xi}_n^{cp,t}|| - \sqrt{\frac{2}{3}}\left(R_0(\theta_n,p_n) + R_{n-1}\right)$$

- **If** $F < 0$: *elastic time step*

    - *Plastic multiplier:* $\lambda = 0$

- **Else**: *plastic time step*

    - *Compute plastic multiplier numerically (using the above abbreviations)*

$$\left|\sqrt{\frac{2}{3}}(R_0 + R_n(\gamma_n)) + g_2(\gamma_n)\tau_n\gamma_n\right| = ||\boldsymbol{G}_1(\gamma_n)||$$

- *Update plastic and TRIP quantities, if necessary:*

    Use (A.9) to calculate first $||\boldsymbol{\xi}_n^{cp}||$ and then $\boldsymbol{\xi}_n^{cp}$ via:

$$\boldsymbol{\xi}_n^{cp}\left(1 + (2\mu + \frac{2}{3}d_x\hat{c} - df_2)\tau_n\gamma_n\frac{1}{||\boldsymbol{\xi}_n^{cp}||}\right) = \boldsymbol{\sigma}_n^{t*} - d_x\boldsymbol{X}_{n-1}^{cp} - d\boldsymbol{F}_1$$

$$\boldsymbol{\xi}_n^{up} = \frac{1}{1 + f_3\tau_n\kappa\left(\sum_i^N \dot{\Phi}_i(p_i)\dot{p}_i\right)}\left(\boldsymbol{F}_1 - f_2\tau_n\gamma_n\frac{\boldsymbol{\xi}_n^{cp}}{||\boldsymbol{\xi}_n^{cp}||}\right)$$

$$\boldsymbol{\sigma}_n^{c*} = \boldsymbol{\sigma}_n^{t*} - 2\mu\tau_n\gamma_n\frac{\boldsymbol{\xi}_n^{cp}}{||\boldsymbol{\xi}_n^{cp}||} - 3\mu\tau_n\kappa\boldsymbol{\xi}_n^{up}\sum_i^N \dot{\Phi}_i(p_i)\dot{p}_i$$

$$X_n^{cp} = \frac{1}{1 + \sqrt{\frac{2}{3}}\gamma_n \hat{b}\tau_n} \left(X_{n-1}^{cp} + \frac{2}{3}\tau_n c\gamma_n \frac{\xi_n^{cp}}{||\xi_n^{cp}||} + \right.$$

$$\left. + c_{int}\tau_n \kappa \xi_n^{up} \sum_i^N \dot{\Phi}_i(p_i)\dot{p}_i \right)$$

$$X_n^{up} = X_{n-1}^{up} + \frac{2}{3}\tau_n c_{int}\gamma_n \frac{\xi_n^{cp}}{||\xi_n^{cp}||} - \sqrt{\frac{2}{3}}\gamma_n c_{int} \frac{\hat{b}}{\hat{c}}\tau_n X_n^{cp} +$$

$$+ \tau_n c_{up}\kappa \xi_n^{up} \sum_i^N \dot{\Phi}_i(p_i)\dot{p}_i$$

$$\epsilon_n^{cp} = \epsilon_{n-1}^{cp} + \tau_n \gamma_n \frac{\xi_n^{cp}}{||\xi_n^{cp}||}$$

$$s_n^{cp} = s_{n-1}^{cp} + \tau_n \sqrt{\frac{2}{3}}\gamma_n$$

$$R_n = \sum_i p_i R_n^i \qquad R_n^i = \frac{1}{1 + \hat{\beta}_i \tau_n \sqrt{\frac{2}{3}}\gamma_n} \left(R_{n-1}^i + \hat{\gamma}_i \tau_n \sqrt{\frac{2}{3}}\gamma_n \right)$$

$$\epsilon_n^{up} = \epsilon_{n-1}^{up} + \frac{3}{2}\tau_n \kappa \xi_n^{up} \sum_i^N \dot{\Phi}_i(p_i)\dot{p}_i$$

Remark. As already mentioned in Section 3.5 the coupling between classical plasticity and transformation induced plasticity can make it necessary to update of the plastic quantities even in elastic time steps.

APPENDIX

B

List of symbols and used material parameters

In this chapter will first give a list of symbols and afterwards specify all used material parameters.

θ temperature

c_e specific heat

κ heat conductivity

L latent heat

P vector of phases

δ heat transfer coefficient

θ_{ext} external temperature

θ_{ms} martensite start temperature

θ_{mf} martensite finish temperature

θ_{m0} parameter of the Koistinen-Marburger model

$\boldsymbol{u}$	displacement
$\boldsymbol{\sigma}$	stress tensor
$\boldsymbol{f}$	body force
$\boldsymbol{\epsilon}$	strain tensor
$\boldsymbol{\epsilon}^{el,cp,up}$	elastic, classic plastic, TRIP part of the strain tensor
μ, λ	Lamé coefficients
E	Young modulus
ν	Poisson ration
K	bulk modulus
ρ	density
F	yield function
$\boldsymbol{X}^{cp}$	back stress of classical plasticity
R_0	yield stress
R	increase of the radius of the yield surface
γ	plastic multiplier
s_{cp}	accumulated plastic strain
$\boldsymbol{X}^{up}$	back stress of transformation induced plasticity
c, m, c_{cp}	parameters of the Ramberg-Osgood and Prager model
$\hat{c}, \hat{b}, \hat{\gamma}, \hat{\beta}$	parameters of the Armstrong-Fredrick model
c_{up}, c_{int}	parameters for back stress coupling
κ_{up}	Greenwood-Johnson parameter
Φ	saturation function of transformation induced plasticity

Throughout this thesis all material parameters are taken from [ADF+08a, ADF+08b] if not specified otherwise. For the convenience of the reader, we will also repeat the material parameters here.

All quantities in the table below depend on the temperature and on the phase fractions of austenite and martensite. We assume a linear mixture rule between both phases, and specify for each phase the polynomial coefficients of the corresponding quantity, e.g. $\kappa(\theta, a, m) = \kappa_a(\theta) \cdot a + \kappa_m(\theta) \cdot m$, where κ_a, κ_m are polynomials in θ with coefficients p_i, e.g. $\kappa_m(\theta) = \sum p_i \theta^i$.

Note that the yield stress is defined as $R_0(\theta, a, m) = R_{0a}(\theta) \cdot a + R_{0m}(\theta) \cdot m$, where $R_{0a} = K_a(\theta) \cdot (0.00005)^{n_a(\theta)}$.

		p_0	p_1	p_2	p_3
heat conductivity	M	$4.355 \cdot 10^{-2}$	$-1.6 \cdot 10^{-6}$	$-4.18 \cdot 10^{-8}$	$1.82 \cdot 10^{11}$
$\kappa \left[\frac{W}{mmC}\right]$	A	$1.46 \cdot 10^{-2}$	$1.27 \cdot 10^{-5}$		
specific heat	M	$4.22 \cdot 10^2$	$9.31 \cdot 10^{-1}$	$-2.14 \cdot 10^{-3}$	$2.64 \cdot 10^{-6}$
$c_e \left[\frac{J}{kgC}\right]$	A	$4.54 \cdot 10^2$	$3.88 \cdot 10^{-1}$	$-3.22 \cdot 10^{-4}$	$1.1 \cdot 10^{-7}$
density	M	7770	$-2.53 \cdot 10^{-1}$		
$\rho \left[\frac{kg}{m^3}\right]$	A	8041.4	$-5.74 \cdot 10^{-1}$	$2.6 \cdot 10^{-5}$	
Young modulus	M	214240	-82.85		
$E[MPa]$	A	266930	221.3		
Poisson ratio	M	0.344	10^{-4}		
$\nu[-]$	A	0.233	$2.5 \cdot 10^{-4}$		
Ramberg-Osgood	M	5724.1	-6.76		
$c[MPa]$	A	960.7	-1.034		
Ramberg-Osgood	M	0.1278	$8.449 \cdot 10^{-5}$		
$m[-]$	A	0.1092			
yield stress					
$K[MPa]$	M	5724.1	-6.76		
	A	960.7	-1.034		
$n[-]$	M	0.1278	$8.449 \cdot 10^{-5}$		
	A	0.1092			

The latent heat is given by the difference of the enthalpy of austenite and martensite: $L = e_a - e_m$, where e_a, e_m are again polynomials in θ with coefficients p_i, e.g. $e_m(\theta) = \sum p_i \theta^i$.

		p_0	p_1	p_2	p_3	p_4
enthalpy	M	$3.82 \cdot 10^3$	$4.22 \cdot 10^2$	$4.66 \cdot 10^{-1}$	$-7.13 \cdot 10^{-4}$	$6.6 \cdot 10^{-7}$
$e \left[\frac{J}{kg}\right]$	A	$8.18 \cdot 10^4$	$4.54 \cdot 10^2$	$1.94 \cdot 10^1$	$-1.07 \cdot 10^{-4}$	$2.75 \cdot 10^{-8}$

martensite start temperature	$\theta_{ms} = 211°C$
martensite finish temperature	$\theta_{mf} = -174°C$
Koistinen-Marburger	$\theta_{m0} = 93.4[C]$
Greenwood-Johnson parameter	$\kappa_{up} = 7 \cdot 10^{-5} \left[\frac{1}{MPa}\right]$

Bibliography

[ACZ99]   Alberty, J., C. Carstensen, and D. Zarrabi: *Adaptive numerical analysis in primal elastoplasticity with hardening.* Computer methods in applied mechanics and engineering, 171:175–204, 1999.

[ADF$^+$08a] Acht, C., M. Dalgic, F. Frerichs, M. Hunkel, A. Irretier, T. Lübben, and H. Surm: *Ermittlung der Materialdaten zur Simulation des Durchhärtens von Komponenten aus 100Cr6 - Teil 1.* HTM Journal of Heat Treatment and Materials, 63:234–244, 2008.

[ADF$^+$08b] Acht, C., M. Dalgic, F. Frerichs, M. Hunkel, A. Irretier, T. Lübben, and H. Surm: *Ermittlung der Materialdaten zur Simulation des Durchhärtens von Komponenten aus 100Cr6 - Teil 2.* HTM Journal of Heat Treatment and Materials, 63:362–371, 2008.

[AF07]   Armstrong, P.J. and C.O. Frederick: *A mathematical representation of the multiaxial bauschinger effect.* Materials at high temperatures, 24(1):1–26, 2007. Reprint of 1966: CEGB Report No. RD/B/N 731, Berkeley, UK.

[Ahr03]   Ahrens, U.: *Beanspruchungsabhängiges Umwandlungsverhalten und Umwandlungsplastizität niedrig legierter Stähle mit unterschiedlich hohen Kohlenstoffgehalten.* PhD thesis, Universtät Paderborn, 2003.

[Bet93]   Betten, J.: *Kontinuumsmechanik – Elasto-, Plasto- und Kriechmechanik.* Springer, 1993.

[BG02]   Becker, W. and D. Gross: *Mechanik elastischer Körper und Strukturen.* Springer, 2002.

[BHSW04]   Böhm, M., M. Hunkel, A. Schmidt, and M. Wolff: *Evaluation of various phasetransition models for 100Cr6 for application in commercial FEM programs.* Journal de Physique IV France, 120:581–589, 2004.

[Böt07]   Böttcher, S.: *Zur mathematischen Aufgabe der Thermoelastizität unter Berücksichtigung von Phasenumwandlungen und Umwandlungsplastizität.* Diplomarbeit, Universität Bremen, Zentrum für Technomathematik, 2007.

[Bra07]   Braess, D.: *Finite Elemente - Theorie, schnelle Löser und Anwendungen in der Elastizitätstheorie.* Springer, 2007.

[CA03]   Carstensen, C. and J. Alberty: *Averaging techniques for reliable a posteriori FE-error control in elastoplasticity with hardening.* Computer methods in applied mechanics and engineering, 192:1435–1450, 2003.

[Cha08]   Chaboche, J.L.: *A review of some plasticity and viscoplasticity constitutive theories.* International Journal of Plasticity, 24:1642–1693, 2008.

[COV06]   Carstensen, C., A. Orlando, and J. Valdman: *A convergent adaptive finite element method for the primal problem of elastoplasticity.* International Journal for numerical Methods in Engineering, 67:1851–1887, 2006.

[CR06]   Chełmiński, K. and R. Racke: *Mathematical analysis of thermoplasticity with linear kinematic hardening.* Journal of Applied Analysis, 1(12):037–057, 2006.

[CS95]   Cailletaud, G. and K. Saï: *Study of plastic/viscoplastic models with various inelastic mechanisms.* International Journal of Plasticity, 11:991–1005, 1995.

[DIL+09]   Dalgic, M., A. Irretier, T. Lübben, C. Simsir, and H. W. Zoch: *Fortschritte der Kennwertermittlung für Forschung und Praxis*, chapter Untersuchung des zyklischen Verhaltens des metastabilen Austenits im Wälzlagerwerkstoff 100Cr6, pages 169–176. Verlag Stahleisen GmbH, 2009.

[DLRW06] Djoko, J.K., B.P. Lamichhane, B.D. Reddy, and B.I. Wohlmuth: *Conditions for equivalence between the Hu-Washizu and related formulations, and computational behaviour in the incompressible limit.* Computer methods in applied mechanics and engineering, 195:4161–4178, 2006.

[DLZ06]   Dalgic, M., G. Löwisch, and H. W. Zoch: *Beschreibung der Umwandlungsplastizität auf Grund innerer Spannungen während der Phasentransformation des Stahls 100Cr6.* HTM Journal of Heat Treatment and Materials, 61(4):222–228, 2006.

[Dog93]   Doghri, I.: *Fully implicit integration and consistent tangent modulus in elastoplasticity.* International Journal for numerical Methods in Engineering, 36:3915–3932, 1993.

[dSNPO08] Souza Neto, E.A. de, D. Perić, and D.R.J. Owen: *Computational methods for plasticity - Theory and application.* John Wiley & Sons, Inc., 2008.

[EJ91]   Erikkson, K. and C. Johnson: *Adaptive finite element methods for parabolic problems I: a linear model problem.* SIAM Journal on Numerical Analysis, 28(1):43–77, 1991.

[FLHZ09] Frerichs, F., T. Lüben, F. Hoffmann, and H. W. Zoch: *Distortion of conical shaped bearing rings made of SAE 52100.* Materialwissenschaft und Werkstofftechnik, 5-6:402–407, 2009.

[FLS91]   Feynman, R.P., R.B. Leighton, and M. Sands: *Feynman Vorlesungen über Physik Band II: Elektromagnetismus und Struktur der Materie.* Oldenbourg Verlag, 1991.

[Gei04]   Geilenkothen, A.: *Efficient Solvers and Error Estimates for a Mixed Method in Elastoplasticity.* PhD thesis, Universität Hannover, 2004.

[GJ65]   Greenwood, G.W. and R.H. Johnson: *The deformation of metals under small stresses during phase transformations.* Proceedings of the Royal Society London 283A, pages 403–422, 1965.

[GR92]   Großmann, Ch. and H. G. Roos: *Numerik partieller Differentialgleichungen.* Teubner, 1992.

[Hau02]   Haupt, P.: *Continuum Mechanics and Theory of Materials.* Springer, 2002.

[Hel98]   Helm, D.: *Experimentelle Untersuchung und phänomenologische Model-
            lierung thermomechanischer Kopplungseffekte in der Metallplastizität.* In
            Berichte des Institus für technische Mechanik der Universität Kassel. S. Hart-
            man and C. Tsamakis, 1998.

[HFP+07]   Hunkel, M., F. Frerichs, C. Prinz, H. Surm, F. Hoffman, and H. W. Zoch:
            *Size change due to anisotropic dilatation behaviour of a low alloy SAE 51200
            steel.* Steel Research International, 78(1):45–51, 2007.

[HH93]   Hartmann, S. and P. Haupt: *Stress computation and consistent tangent oper-
            ator using non-linear kinematic hardening models.* International Journal for
            numerical Methods in Engineering, 36:3801–3814, 1993.

[Höm96]   Hömberg, D.: *A numerical simulation of the Jominy end-quench test.* Acta
            Materialia, 11:4375–4385, 1996.

[Höm97]   Hömberg, D.: *Irreversible phase transitions in steel.* Mathematical Methods
            is the applied sciences, 20:59–77, 1997.

[Höm04]   Hömberg, D.: *A mathematical model for induction hardening including me-
            chanical effects.* Nonlinear Analysis: Real World Applications, 5:55–90, 2004.

[Hor92]   Horstmann, D.: *Das Zustandsschaubild Eisen-Kohlenstoff und die Grund-
            lagen der Wärmebehandlung der Eisen-Kohlenstoff-Legierungen.* Verlag
            Stahleisen, 1992.

[HPS92]   Hartley, P., I. Phillinger, and C. Sturges (editors): *Numerical Modelling of
            Material Deformation Processes – Research, Development and Applications.*
            Springer, 1992.

[HR99]   Han, W. and B.D. Reddy: *Plasticity: Mathematical Theory and Numerical
            Analysis.* Interdisciplinary applied mathematics: mechanics and materials.
            Springer, 1999.

[Hun]   Hunkel, M.: *Presentation in an internal reaserch group.* To be published
            soon.

[Hüß07]   Hüßler, I.: *Mathematische Untersuchungen eines gekoppelten Systems von
            ODE und PDE zur Modellierung von Phasenumwandlungen im Stahl.* Di-
            plomarbeit, Universität Bremen, Zentrum für Technomathematik, 2007.

[INKM81]   Inoue, T., S. Nagaki, T. Kishino, and M. Monkawa: *Description of tranfor-mation kinetics, heat conduction and elastic-plastic stress in the course of quenching and tempering of some steels.* Ingenieur-Archiv, 50:315–327, 1981.

[Joh76]   Johnson, C.: *Existence theorems for plasticity problems.*   Journal de mathématiques pures et appliquées, 55:431–444, 1976.

[Joh78]   Johnson, C.: *On plasticity with hardening.* Journal of mathematical Analysis and Application, 62:325–336, 1978.

[KM59]   Koistinen, D.P. and R.E: Marburger: *A gerneral equation prescribing the extent of the austenite-martensite transformation in pure iron-carbon and plain carbon steels.* Acta Metallurgica, 7(1):59–60, 1959.

[LD84]   Leblond, J.B. and J. Devaux: *A new kinetic model for anisothermal metal-lurgical transformations in steels including effect of austenite grain size.* Acta Metallurgica, 32:137–146, 1984.

[LMD86a]   Leblond, J.B., G. Mottet, and J.C. Devaux: *A theoretical and numerical approach to the plastic behavior of steels during phase transformations - I.* Journal of the Mechanics and Physics of Solids, 34:395–409, 1986.

[LMD86b]   Leblond, J.B., G. Mottet, and J.C. Devaux: *A theoretical and numerical approach to the plastic behavior of steels during phase transformations - II.* Journal of the Mechanics and Physics of Solids, 34:411–432, 1986.

[LMD89a]   Leblond, J.B., G. Mottet, and J.C. Devaux: *Mathematical modelling of trans-formation plasticity in steels. I: Case of ideal-plastic phases.* International Journal of Plasticity, 5:551 – 572, 1989.

[LMD89b]   Leblond, J.B., G. Mottet, and J.C. Devaux: *Mathematical modelling of trans-formation plasticity in steels. II: Coupling with strain hardening phenomena.* International Journal of Plasticity, 5:573 – 591, 1989.

[Mac92]   Macherauch, E.: *Praktikum in Werkstoffkunde.* Vieweg, 1992.

[Mag66]   Magee, C.L.: *Transformation kinetics, micro-plasticity and aging of marten-site in Fe-31Ni.* PhD thesis, Carnegie Mellon Institute of Technology, Pitts-burg, 1966.

[Mag68]   Magee, C.L.: *The nucleation of martensite.* In *Phase transformations - Papers presented at a Seminar of the American Society for Metals*, Metals Park, Ohio, 1968. American Society for Metals.

[Mah99]   Mahnken, R.: *Improved implementation of an algorithm for non-linear isotropic/kinematic hardening in elastoplasticity.* Communications in numerical methods in engineering, 15:745–754, 1999.

[Mah04]   Mahnken, R.: *Encyclopedia of Computational Mechanics: Identification of Material Parameters for Constitutive Equations*, volume 2, chapter 19, pages 637–655. John Wiley & Sons, Inc., 2004.

[Max61]   Max-Planck-Institut für Eisenforschung in Zusammenarbeit mit dem Werkstoffausschuss des Vereins Deutscher Eisenhütten (editor): *Atlas zur Wärmebehandlung der Stähle.* Verlag Stahleisen, 1961.

[MBTS07]   Meftah, S., F. Barbe, L. Taleb, and F. Sidoroff: *Parametric numerical simulations of TRIP and its interaction with classical plasticity in martensitic transformation.* European Journal of Mechanics A/Solids, 26:688–700, 2007.

[MNS]   Morin, P., R.H. Nochetto, and K.G. Siebert: *Basic principles for convergence of adaptive higher-order FEM-application to linear elasticity.* in preparation.

[MSA09]   Mahnken, R., A. Schneidt, and T. Antretter: *Macro modelling and homogenization for transformation induced plasticity of low-allow steel.* International Journal of Plasticity, 25:183–204, 2009.

[Pra49]   Prager, W.: *Recent developments in the mathematical theory of plasticity.* Journal of Applied Physics, 20 (3):235–241, 1949.

[Pra56]   Prager, W.: *A new method of analyzing stresses and strains in work hardening plastic solids.* Journal of Applied Mechanics, 23:493–496, 1956.

[QV97]   Quarteroni, A. and A. Valli: *Numerical Approximation of Partial Differential Equations.* Number 23 in *Springer Series in Computational Mathematics*. Springer, 1997.

[RO43]   Ramberg, W. and W.R. Osgood: *Description of stress-strain curves by three parameters.* Technical Report 902, NACA Technical Note, 1943.

[Sch83]   Schwendemann, H.: *Die thermische Restaustenitstabilisierung bei den Stählen 100Cr6 und X210Cr12.* PhD thesis, University of Karlsruhe, 1983.

[SD96]   Stouffer, D.C. and L.T. Dame: *Inelastic deformations in metals – Models,*
            mechanical properties and metallurgy. John Wiley & Sons, Inc., 1996.

[ŞDL⁺10]   Şimşir, C., M. Dalgiç, T. Lübben, A. Irretier, M. Wolff, and H. W. Zoch: *The*
            Bauschinger effect in supercooled austenite of SAE 52100 steel. submitted to
            Acta Materialia, 2010.

[SFHW09]   Suhr, B., F. Frerichs, I. Hüßler, and M. Wolff: *Evaluation of models for*
            martensitic transformation and TRIP via comparison of experiments and
            simulations. Materialwissenschaft und Werkstofftechnik, 40:460–465, 2009.

[SH97]   Simo, J.C. and T.J.R. Hughes: *Computational Inelasticity.* Interdisciplinary
            applied mathematics: mechanics and materials. Springer, 1997.

[Sho96]   Showalter, R.E.: *Monotone Operators in Banach Space and Nonlinear Partial*
            Differential Equations, volume 49 of *Mathematical Surveys and Monographs.*
            American Mathematical Society, 1996.

[SS05]   Schmidt, A. and K.G. Siebert: *Design of Adaptive Finite Element Software*
            - The Finite Element Toolox ALBERTA. Number 42 in *Lecture Notes in*
            Computational Science and Engineering. Springer, 2005.

[SSBH07]   Schmidt, A., B. Suhr, M. Böhm, and M. Hunkel: *Simulation of anisotropic*
            dilation behaviour of layered and banded steel samples during phase changes.
            Philosophical Magazine Letters, 87(11):871–881, 2007.

[ST85]   Simo, J.C. and R.L. Taylor: *Consistent tangent operators for rate independent*
            elastoplasticity. Computer methods in applied mechanics and engineering,
            48(1):101–118, 1985.

[SWB03]   Schmidt, A., M. Wolff, and M. Böhm: *Numerische Untersuchungen*
            für ein Modell des Materialverhaltens mit Umwandlungsplastizität und
            Phasenumwandlungen beim Stahl 100Cr6. Technical report, Universität Bre-
            men, Berichte aus der Technomathematik, 2003.

[TC10]   Taleb, L. and G. Cailletaud: *An updated version of the multimechanism model*
            for cyclic plasticity. International Journal of Plasticity (Article in press),
            2010.

[TP02]   Taleb, L. and S. Petit: *Elastoplasticity and phase transformations in ferrous*
            alloys: some discrepencies between experiments and modeling. Journal de
            Physique IV France, 12:11–187, 2002.

[TPG06]   Taleb, L. and S. Petit-Grostabussiat: *New investigations on transformation
           induced plasticity and its interaction with classical plasticity.* International
           Journal of Plasticity, 22:110–130, 2006.

[Ver96]   Verfürth, R.: *A review of a posteriori error estimation and adaptive mesh-
           refinement techniques.* Wiley Teubner, 1996.

[WB06]   Wolff, M. and M. Böhm: *Transformation-induced plasticity in steel - general
           modelling, analysis and parameter indentification.* Technical Report 06-02,
           Universität Bremen, Berichte aus der Technomathematik, 2006.

[WBB07]   Wollf, M., M. Böhm, and S. Böttcher: *Phase transformations in steel in the
           multi-phase case – gerneral modelling and parameter identification.* Techni-
           cal Report 07-02, Universität Bremen, Berichte aus der Technomathematik,
           2007.

[WBBL07]   Wolff, M., S. Boettcher, M. Böhm, and I. Loresch: *Vergleichende Be-
           wertung von makroskopischen Modellen für die austenitisch-perlitische
           Phasenumwandlung im Stahl 100Cr6.* Technical Report 07-03, Universität
           Bremen, Berichte aus der Technomathematik, 2007.

[WBD$^+$06] Wollf, M., M. Böhm, M. Dalgic, G. Löwisch, N. Lysenko, and J. Rath: *Pa-
           rameter identification for a TRIP model with backstress.* Computational Ma-
           terials Science, 37:37–41, 2006.

[WBD$^+$07] Wolff, M., M. Böhm, M. Dalgic, G. Löwisch, and J. Rath: *Validation of
           a TRIP model with backstress for the pearlitic transformation of the steel
           100Cr6 under step-wise loads.* Computational Materials Sciences, 39:49–54,
           2007.

[WBDH08]   Wolff, M., M. Böhm, M. Dalgic, and I. Hüßler: *Evaluation of models for TRIP
           and stress-dependend transformation behaviour for the martensitic transfor-
           mation of the steel 100Cr6.* Computational Materials Sciences, 43:108 – 114,
           2008.

[WBH08]   Wolff, M., M. Böhm, and D. Helm: *Material behavior of steel - modelling of
           complex phenomena and thermodynamic consistency.* International Journal
           of Plasticity, 24:746–774, 2008.

[WBMS]   Wollf, M., M. Böhm, R. Mahnken, and B. Suhr: *Implementation of an algorithm for general material behaviour of steel taking interaction of plasticity and transformation induced plasticity into account.* To appear.

[WBS09]   Wolff, M., M. Böhm, and B. Suhr: *Comparison of different approaches to transformation-induced plasticity in steel.* Materialwissenschaft und Werkstofftechnik, 40:454–459, 2009.

[Weg04]   Wegst, C. (editor): *Stahlschlüssel Taschenbuch.* Stahlschlüssel, 2004.

[Wol08]   Wolff, M.: *Zur Modellierung des Materialverhaltens von Stahl unter Berücksichtigung von Phasenumwandlung und Umwandlungsplastizität.* Habilitationsschrift. University of Bremen, 2008.

[Yu97]   Yu, H.Y.: *A new model for the volume fraction of martensitic transformations.* Metallurgical and Materials Transactions, 28A:2499–2506, 1997.